Climate Change, Uncertainty and Agriculture

Abouth the Authors

Prof. (Dr.) Sankar Kr Acharya, presently, Dean, Post Graduate Studies; former Head, Dept. of Agril Extension and Director, Extension Education, Bidhan Chandra Krishi Viswavidyalaya, Mohanpur, WB, born on 6th October, 1960, started his career as Assistant Professor at BCKV in 1988 and has been in teaching, research and extension over 35 years. An erudite teacher as well as an elegant speaker, is internationally acclaimed for his unique research domain of Social Entropy and Energy Metabolism, Social Ecology and Environmental Sociology, Enterprise Ecology Framework, Farm Stewardships: The Transition from Conservation, Technology Socialization Process, and The Social Ecology of Livelihood, which have got tremendous policy as well as scholastic importance. Research Publication: 247 in National and International Journals. Book Publication: 108 books authored / co-authored, covering areas Participatory Extension Planning and Management; Value, Extension Strategy for Natural Resource Management; Social Metabolism, Farm Energy Metabolism, Community Perception on Climate Change , Social Entropy and Chaos, Decision ecology; Expert Member, Agriculture Commission; Marketing and Extension Sub Committee, Govt. of West Bengal; Expert member WWF projects in BTR (Buxa Tiger Reserve Project); Expert member; DFID project on Primary Education (DPEP); Rapid Environment Impact Analysis, Visited Italy, France, Germany, China , Sri Lanka and Bangladesh. Delivered 31 Key note addresses, 13, invited and lead speakers in International Congress, Chaired 19 Sessions in International conferences, and Editors of a score of national and international journals. He has so far been awarded with 15 best paper awards in National and International Conferences Organizing Secretary of two International Conferences, held in the years, 2016 & 2018, Jointly Organized by Department of Agricultural Extension, BCKV, WB and Krishisanskriti, New Delhi. He has also been honoured to be selected as the Convener of Panel (PE-32) entitled The hunger, poverty and silence, of IUAES, University of Manchester, UK, 2013 .Co-PI of ICAR, NAHEP and World Bank funded project on Conservation Agriculture. So, far he has successfully guided 19 Ph D scholars, of which one awarded with DAAD fellowship, Germany and two other placed as visiting scholars at Ohio State University and Cornel University, 2020-21, USA; 83 M Sc students from BCKV CU, KU, Vidyasagar University. He is the fellow of ISEE, IARI, New Delhi, Fellow, BIOVED, He is now acting as editor, reviewer of a good number of national and international journals. Fellow Award, ISEE(ICAR-IARI, New Delhi), 2019Awards and Distinctions: Distinguish Scholar Award, Krishisanskriti, New Delhi; Certificate of Merit for standing First class first at M Sc(Ag) in Agricultural Extension, 1986; ;Honorary Appointment to the Research Board of Advisors, The American Biographic Institute (2003). He has been honored to be selected as the

Convener of Panel (PE-32) entitled the hunger, poverty and silence, of World Congress, IUAES, University of Manchester, UK, 2013 ;He has been honored to be selected as the Convener of Panel (P-102) entitled ,' Uncertainties, Unpredictability and Marginalization: The impact and mitigation for survival of agriculture and humanity, of the 19th World Congress, IUAES-WAU, 14-20 October, 2023, New Delhi, India, 2023 ;Honored to be nominated as Expert Member, Working group on Agricultural Extension and Training, West Bengal State Agriculture Commission (2007) ;Honored to be nominated by the Hon'ble Vice-Chancellor, BCKV to Act as an Expert Member, DPEP, A DFID, UK, project for planning and revamping primary education through POA of GOVT of India.(1995) Expert Member, Buxa Tiger Reserve Project funded by WWF(World Wide Wild Life Funding), 1998 ;Distinguished Professor and Academician Award, Research Education Solution (RES), M S Swaminathan School of Agriculture, Centurion University(2022); Diamond Achievers Award, Council for Academic Performance and Appraisal, New Delhi (2022); Honorary Fellowship Award, BIOVED Research Institute of Agriculture and Technology, Prayagraj on Occasion of Agricultural Scientists and Farmers' Congress.Achievers of German patent (Utility model no.20202310273) :Novel LOT based inventory management system with radio controlled pallet racking for storage solution.

Partha Saren, born in Hooghly in 1999, is currently pursuing Post Graduate Diploma in Agribusiness (PGDAB). Presently he is engaged in research works for the dissertation of PGDABdegree and is working on writing other research papers and book chapters as well. He has completed his Bachelor of Science Agriculture (Honours) with First Division with Distinction from Bidhan Chandra Krishi Viswavidyalaya in 2021. He has completed Master of Science (Ag.) in Agricultural Extension Education with First Class from the Department of Agricultural Extension, BCKV in 2023. He has experience of RAWE as an Intern and ELP as a Trainee under the one year Student READY Programme of ICAR. He has worked under renowned Extension Personnel in India, Prof. S. K. Acharya in Knowledge Lab, Department of Agricultural Extension, BCKV as a Post Graduate Researcher to complete his Master's Thesis. Under the guidance of his PG Research Advisor, he has developed keen interest in Data Analytics. He has been trained on Basics of Data Science including Python, Predictive Modelling and Statistical Data Analysis by Internshala, a NSDC approved Edu-Tech giant.

Climate Change, Uncertainty and Agriculture

Sankar Kumar Acharya
Partha Saren

PMW

PMW
FU 75, Pitam Pura
New Delhi 110034
publishmywork23@gmail.com
Mob. 91-8447075807

Print ISBN: 978-81-96458-89-8

ebook ISBN: 978-81-96458-86-7

Preface

Climate change is a complex and dynamic phenomenon that has significant impacts on agriculture, especially in terms of uncertainty and variability. Agricultural dynamics refer to the various factors that affect agricultural production, including climate, soil, water availability, pests, and diseases, as well as market forces and government policies. Climate change has significant impacts on agricultural dynamics. Variations in temperature, patterns of rainfall, and occurrences of extreme weather can influence crop yields and livestock production in both beneficial and detrimental ways (UCAR, 2021). The effect of climate change on farming can differ based on the geographical location and the specific agricultural practices employed. For example, in some areas, increased rainfall may lead to increased crop yields, while in other areas; droughts may lead to crop failures and loss of income for farmers. Additionally, rising temperatures can cause reduced crop yields and decreased quality of some crops. Uncertainty is a term used to describe a situation where there is a lack of information, knowledge, or predictability about a particular outcome or event. In agricultural dynamics, uncertainty can refer to a range of factors that can affect the production, distribution, and consumption of agricultural products. One of the most significant sources of uncertainty in agriculture is the weather. The productivity of crops can be significantly influenced by weather conditions, including rainfall and temperature and farmers need to effectively adjust their agricultural practices to adapt to these changing weather patterns The social, ecological and economic interpretation of climate change by farmers involves understanding the interactions between the natural environment, social systems, and economic factors (Spash, Clive L, 2011). Farmers need to consider these interactions when making decisions about their farming practices. For example, changes in climate can affect the availability of water, soil fertility, and crop yields, which can in turn affect social and economic factors such as food security and income. Farmers need to take into account the wider consequences of their agricultural practices on society and the environment which includes considering the impacts on biodiversity, ecosystems, and human well-being (Warsaw et al., 2021). The book, A Real Time Interpretation of Climate Change, Uncertainty and Agricultural Dynamics, presents an excellent piece of empirical study with application of Fuzzy Cognitive Modelling to display and elucidate differential perceptions and scenario of climate change and its impacts on agriculture and livelihood. I hope and believe, this book is of huge utility to the faculties, policy makers, scholars beyond the fences of discipline and for the global audience as well.

Authors

Contents

1

Prologue

Climate change is a complex and dynamic phenomenon that has significant impacts on agriculture, especially in terms of uncertainty and variability. Agricultural dynamics refer to the various factors that affect agricultural production, including climate, soil, water availability, pests, and diseases, as well as market forces and government policies. Climate change has significant impacts on agricultural dynamics. Variations in temperature, patterns of rainfall, and occurrences of extreme weather can influence crop yields and livestock production in both beneficial and detrimental ways (UCAR, 2021). The effects of climate change on farming can differ based on the geographical location and the specific agricultural practices employed (Wikipedia Contributors, 2019). For example, in some areas, increased rainfall may lead to increased crop yields, while in other areas, droughts may lead to crop failures and loss of income for farmers. Additionally, rising temperatures can cause reduced crop yields and decreased quality of some crops.

Uncertainty is a term used to describe a situation where there is a lack of information, knowledge, or predictability about a particular outcome or event. In agricultural dynamics, uncertainty can refer to a range of factors that can affect the production, distribution, and consumption of agricultural products. One of the most significant sources of uncertainty in agriculture is the weather. The productivity of crops can be significantly influenced by weather conditions, including rainfall and temperature and farmers need to effectively adjust their agricultural practices to adapt to these changing weather patterns (Guntukulaand Goyari, 2020). For example, droughts or floods can reduce crop yields, while frost or hail can damage crops(Lamaoui et al., 2018).

The social, ecological and economic interpretation of climate change by farmers involves understanding the interactions between the natural environment, social systems, and economic factors (Spash, Clive L, 2011). Farmers need to consider these interactions when making decisions about their farming practices. For example, changes in climate can affect the availability of water, soil fertility, and crop yields, which can in turn affect social and economic factors such as food security and income. Farmers need to take into account the wider consequences of their agricultural practices on society and the environment which includes considering the impacts on biodiversity, ecosystems, and human well-being (Warsaw et al., 2021).

Farmers face various challenges in adapting to these changes and making decisions about their farming practices. To address these issues, a mental modelling approach can be used to understand how farmers interpret and respond to the social, ecological, and economic aspects of climate change. A mental model refers to a cognitive framework that individuals use to comprehend and make decisions regarding intricate systems within the world (Wikipedia Contributors, 2019). Mental modelling involves eliciting the mental models of individuals or groups, which can be used to explore the decision-making processes that underlie their behaviour. This approach has been utilized in the context of climate change and agriculture to gain insights into farmers' perspectives on climate change, uncertainty, and the dynamics of agriculture (Farming First, n.d.).

The mental modelling approach can be used to elicit farmers' mental models of climate change and agriculture, and to explore the factors that influence their decision-making processes. This can offer valuable insights into how farmers comprehend and address uncertainties and variabilities within the realm of climate change (Shipman, 2020). The approach can also be used to identify potential barriers to adaptation and to explore strategies for overcoming these barriers. The mental modelling approach can provide a useful tool for understanding the complex interactions between climate change, agriculture, and society. Through comprehending the mental frameworks of farmers, we can enhance our ability to assist their decision-making procedures and promote more efficient adaptation to the challenges posed by climate change (Öhlmér, 1998).

1.1 Rationale of the Study

The study of climate change, uncertainty, and agricultural dynamics holds significant importance due to the pivotal role of agriculture in the global economy, serving as a crucial sector that sustains the livelihoods of millions and supplies food to people all over the world (Stecula and Merkley, 2019). Climate change has the potential to significantly impact agricultural production, and there is a need to understand how farmers interpret and respond to these changes. This study aims to explore the social, ecological, and economic interpretations of climate change and uncertainty by farmers, using a mental modelling approach. By exploring the mental models that farmers use to understand climate change and its impacts on their agricultural practices, this research can offer valuable perspectives on developing strategies that are more efficient in adapting to and mitigating the effects of climate change (BBC Bitesize, 2012). By comprehending the mental frameworks of farmers, it becomes possible to recognize areas where there is a lack of knowledge and use this insight to shape customized educational and outreach initiatives that cater to the unique requirements of farmers. Moreover, this research can aid in enhancing our comprehension of the intricate interplay among social, ecological, and economic elements that impact agricultural dynamics amidst climate change and uncertainty. Ultimately, the outcome of this investigation can contribute to inform policy and decision-making procedures concerning agriculture, climate change, and the pursuit of sustainability (Pineo et al., 2020). Under such context, needs for the present research are pointed out as follows -

i. *Understanding the impact of climate change on agriculture*: Climate change is expected to have a significant impact on agriculture with changes in temperature and precipitation patterns affecting crop yields and livestock production (PIB Delhi, 2022; Gornall et al., 2010). By studying how farmers perceive and respond to climate change, we can better understand the challenges they face and develop strategies to help them adapt.

ii. *Exploring the role of uncertainty*: Climate change is also associated with a high degree of uncertainty, which can make it difficult for farmers to plan for the future. By exploring how farmers think about uncertainty, we can better understand how they make decisions in the face of uncertain conditions.

iii. *Examining social, ecological, and economic factors*: Agriculture is a complex system that is influenced by a variety of social, ecological, and economic factors. By taking a holistic approach that examines all of these factors together, we can gain a better understanding of how they interact to shape agricultural dynamics (Campbellsville University, 2019).

iv. *Using a mental modelling approach*: Mental modelling is a powerful tool for understanding complex systems, such as agriculture in the face of climate change. By working with farmers to develop mental models of their farming systems, researchers can gain insights into their thought processes and decision-making strategies.

The present study on Climate Change, Uncertainty and Agricultural Dynamics: Social, Ecological and Economical Interpretation by Farmers through Mental Modelling Approach can help us inform more effective strategies for supporting farmers and promoting sustainable agriculture in the face of climate change.

1.2 Research Questions

Climate change is recognized as one of the most significant challenges faced by the agricultural sector worldwide. The complexity and uncertainty of climate change make it difficult for farmers to plan and adapt to changing conditions. This thesis aims to explore how farmers in a particular area comprehend and react to the social, ecological, and economic factors related to climate change by employing a mental modelling methodology. The following research questions are identified to guide this present research based on the needs of the study: a) is it possible to estimate the perception on agricultural change pattern due to climate change and uncertainty? b) What are the variables and is it possible to isolate the variables based on the socio-economic and ecological interpretation by the farmers? c) Is it possible to estimate different levels of functional interactions between the sets of variables under the study out of the analysed data? d) Is it possible to formulate fuzzy logic cognitive map by using mental modelling approach for eliciting farmers' mental models on climate change, uncertainty and agricultural dynamics?

By addressing these research questions, this thesis aims to contribute to a better understanding of the social, ecological, and economic dimensions of climate change in the agricultural sector.

1.3 Objectives

The overall objective of this research is to investigate how farmers in a specific region interpret and respond to the social, ecological, and economic dynamics of climate change and agriculture using a mental modelling approach. The following specific objective statements are pointed out:

The specific objectives of the present study are**(a)**to estimate the farmers' perception on climate change, level of uncertainty and trend of agricultural change for the selected study area;**(b)**to identify and customize the pertinent variables present in the selected study area; **(c)**to elucidate and evaluate the intra and inter level interactions among and between the sets of exogenous variables and consequence variables; **(d)**to formulate a fuzzy logic cognitive map by using mental modelling approach to visualize farmers' perception on climate change, uncertainty and agricultural dynamics; **(e)** to generate micro-sociological policy implications based on the empirical research.

2

Theory and Concepts

In the context of a thesis, theoretical orientation refers to the conceptual or theoretical framework or perspective that informs and shapes the overall approach and analysis of the research (Natalie, 2022). It involves selecting and adopting a specific set of theories, concepts, or models that provide a foundation for understanding and interpreting the research topic. The theoretical orientation helps to guide the methodology, and provide a coherent framework for organizing and analysing the data.

Theoretical orientation in a thesis serves several purposes. It helps to establish a theoretical basis for the study, providing a context for the research and connecting it to existing knowledge and theories in the field. It also helps to guide the selection of appropriate research methods and data analysis techniques. The theoretical orientation shapes the overall structure and content of the thesis, ensuring that the study is grounded in relevant theories and concepts and contributes to the existing body of knowledge. By adopting a specific theoretical orientation, the researcher can provide a clear and systematic framework for understanding and interpreting the findings of the study. This orientation also helps to establish credibility and validity for the research, as it demonstrates the researcher's awareness of and engagement with the existing theoretical perspectives in the field.

This chapter addresses crucial terminology and concepts essential for effectively communicating the ideas of the current study. It lays the groundwork for empirical investigation and aids in the selection of predictor variables. Additionally, this section involves a scholarly and theoretical exercise that influences the operational definition and content of the research. It combines conceptual inputs, axiomatic inputs, logical discussions, and the pursuit of knowledge to construct a theory that aligns with the research objectives and designs.

2.1 Definition of Farmer, Farm, and Farming System

A farmer is an individual or a person who is engaged in the practice of agriculture, specifically the cultivation of crops, raising livestock, or both. Farmers are involved in various activities related to food production, such as planting, tending, and harvesting crops, managing livestock, and maintaining the overall functioning of a farm. Farmers play a vital role in the production of food and other agricultural products. They are responsible for making decisions regarding the selection of crops or livestock, determining the appropriate planting and harvesting times, implementing pest and weed control measures, and managing the overall health and well-being of their animals. Farmers often work long hours and face various challenges, including weather fluctuations, market fluctuations, and the need to adapt to changing technologies and practices in agriculture.

A farm is an area of land, typically rural, that is primarily used for agricultural purposes. It is a place where crops are cultivated, livestock is raised, and other agricultural activities take place. Farms can vary in size and can specialize in different types of agriculture, such as crop farming, dairy farming, poultry farming, or mixed farming. A farm is not only a physical space but also a complex system where various components interact. It typically consists of fields or paddocks where crops are grown, barns or sheds for housing livestock, and infrastructure for storing and processing agricultural products. Farms may also have equipment, such as tractors and machinery, for performing tasks like ploughing, irrigation, and harvesting. Additionally, farms can include support structures like farmhouses, storage facilities, and administrative buildings.

A farming system refers to the combination of practices, techniques, and processes employed by farmers to manage their land, crops, and livestock in a specific agricultural setting. It encompasses the overall approach and strategies used by farmers to optimize production, maintain soil fertility, manage pests and diseases, and ensure sustainable farming practices. A farming system takes into account factors such as climate, soil type, available resources, market demands, and the goals and preferences of the farmer. A farming system is more than just the sum of its individual components. It encompasses the integration of different elements, such as crops, livestock, soil management practices, water usage, and environmental considerations. Farmers may adopt different farming systems based on their location, available resources, and cultural practices. For example, there are conventional farming systems that rely on synthetic fertilizers and pesticides, as well as organic farming systems that prioritize natural inputs and sustainable practices. Agroforestry, aquaculture, and hydroponics are other examples of farming systems that utilize specific techniques and combinations of agricultural activities. Farming systems are influenced by factors such as economic conditions, market demands, societal preferences, and environmental sustainability. The goal is to achieve efficient and sustainable production while minimizing negative impacts on the environment and maximizing social and economic benefits.

Understanding the roles of farmers, farms, and farming systems provides valuable insights into the complexities and importance of agriculture in providing food security, livelihoods, and environmental sustainability.

Fig. 1: Drones are increasingly applied to rationalize the input volume and cost, reducing carbon foot print

2.2 Climate Change and Global Warming

Climate change and global warming are complex and interconnected phenomena that have become significant global concerns. It is essential to establish a theoretical orientation on these issues to grasp their implications, understand their causes, and explore potential solutions.

Climate change refers to long-term alterations in weather patterns, including temperature, precipitation, wind patterns, and other factors that affect the Earth's climate system (United Nations, 2022). Global warming, on the other hand, specifically refers to the increase in Earth's average surface temperature due to the accumulation of greenhouse gases in the atmosphere. These significant alterations have primarily been propelled by human actions, specifically the combustion of non-renewable resources like coal, petroleum, and natural gas (Tabassum, 2023), which result in the emission of carbon dioxide (CO_2) and various other greenhouse gases in large quantities into the atmosphere (Fecht, 2021).

It is important to recognize the gravity of climate change and global warming as they pose significant risks to ecosystems, human societies, and the planet as a whole. The impacts of these phenomena are already being observed, with rising global temperatures leading to widespread consequences. These include the melting of polar ice caps and glaciers, resulting in sea-level rise and increased coastal flooding (Simon, 2021). Extreme weather events such as hurricanes, droughts, and heatwaves have become more frequent and intense, causing severe economic and humanitarian crises (Buis, 2020). Furthermore, climate change disrupts ecosystems, leading to shifts in habitats, biodiversity loss, and negative effects on agriculture, water resources, and public health.

Understanding the theoretical underpinnings of climate change and global warming is crucial in formulating effective policies and strategies to address these issues. By recognizing the scientific consensus and the impacts of climate change, individuals, communities, businesses, and governments can work together to reduce greenhouse gas emissions, transition to sustainable energy sources, protect vulnerable ecosystems, and adapt to the changes already underway.

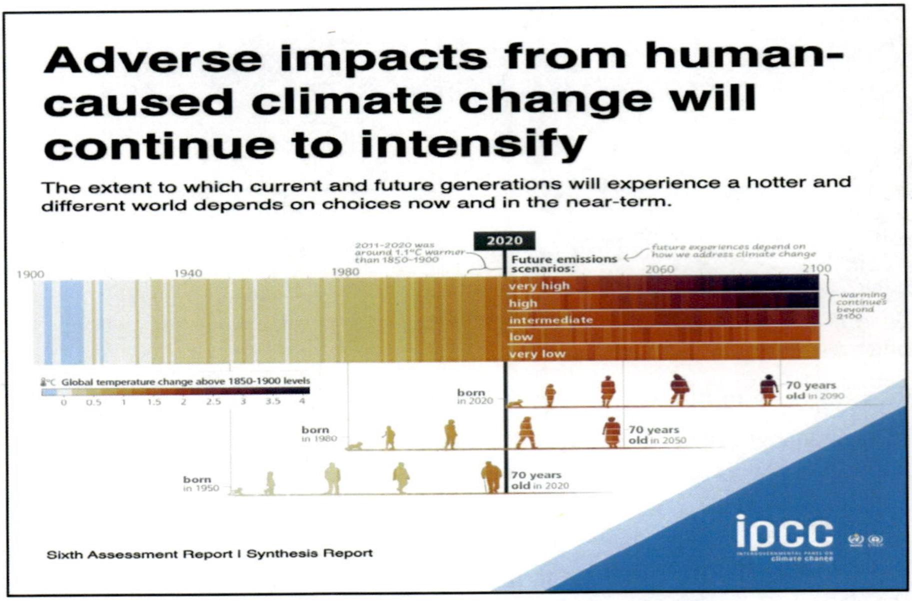

Fig. 2: IPCC predicts more intensified climate change in coming decades.

2.2.1 Scientific Consensus

Scientific consensus refers to the widespread agreement among scientists from various disciplines regarding a particular scientific theory or phenomenon. In the case of climate change and global warming, there is an overwhelming scientific consensus that human activities are the primary drivers of the observed changes in Earth's climate system.

The consensus on climate change is based on extensive research, data analysis, and modelling studies conducted by thousands of scientists worldwide over several decades. These studies draw upon multiple lines of evidence, including measurements of temperature records, atmospheric composition, ice core samples, oceanic data, and historical climate data (Sherwood et al., 2020).

The Intergovernmental Panel on Climate Change (IPCC), a body established by the United Nations, serves as a prominent source of scientific consensus on climate change. The IPCC assesses and synthesizes scientific literature to provide policymakers with comprehensive reports on the state of knowledge regarding climate change (Howarth and Painter, 2016). Its assessments involve the contributions of thousands of scientists who review and evaluate the available evidence to develop a consensus-based understanding of climate change and its impacts.

The scientific consensus on climate change is reflected in numerous scientific organizations and academies worldwide (*Climate Change Science Adequately Reflected in the "Climate Change Sense of the Congress Resolution,"* 2003). Major scientific bodies, including the National Aeronautics and Space Administration (NASA), the National Oceanic and Atmospheric Administration (NOAA), and the World Meteorological Organization (WMO), have all endorsed the consensus position. Additionally, scientific societies such as the American Association for the Advancement of Science (AAAS) and the American Geophysical Union (AGU) have expressed their support for the consensus on climate change (AAAS, 2023; NOAA, 2021; WMO, 2015).

The significance of scientific consensus lies in its role in shaping public understanding, policy decisions, and societal responses to climate change. By acknowledging and acting upon the consensus, policymakers can make informed decisions based on the best available evidence (Pal, 2019). Scientific consensus also plays a vital role in effectively communicating the urgency of climate action to the general public and fostering a collective understanding of the need for mitigation and adaptation measures.

2.2.2 Greenhouse Effect

The greenhouse effect is a natural phenomenon that plays a crucial role in regulating the Earth's temperature and making it suitable for life. It refers to the process by which certain gases in the Earth's atmosphere trap heat from the sun, preventing it from escaping back into space (EnglishClub, 2023). However, human activities have significantly enhanced the greenhouse effect, leading to global warming and climate change.

The greenhouse effect begins with the sun radiating energy in the form of sunlight towards the Earth (British Geological Survey, 2022). About one-third of this incoming solar radiation is reflected back into space by clouds, ice, and the Earth's surface. The remaining two-thirds is absorbed by the Earth's surface, which then emits this energy back into the atmosphere in the form of infrared radiation (heat) (Turgeon and Morse, 2022).

Greenhouse gases, such as carbon dioxide (CO_2), nitrous oxide (N_2O), methane (CH_4) and water vapour, act as a natural blanket in the atmosphere, absorbing and re-emitting the outgoing heat (United States Environmental Protection Agency, 2017). This process keeps the Earth's surface warm enough to sustain life (World 101, 2019). Without the greenhouse effect, the average temperature on Earth would be around -18°C (0°F), rendering the planet inhospitable.

Human activities, particularly the burning of non-renewable resources such as coal, petroleum and natural gas for energy and transportation, have significantly increased the concentration of greenhouse gases in the atmosphere (Vedantu, 2019). Carbon dioxide, released primarily from the combustion of fossil fuels, is the most significant contributor to the enhanced greenhouse effect (US EPA, 2022). Deforestation, which reduces the capacity of forests to absorb CO_2, also contributes to increased greenhouse gas concentrations (Admin, 2017).

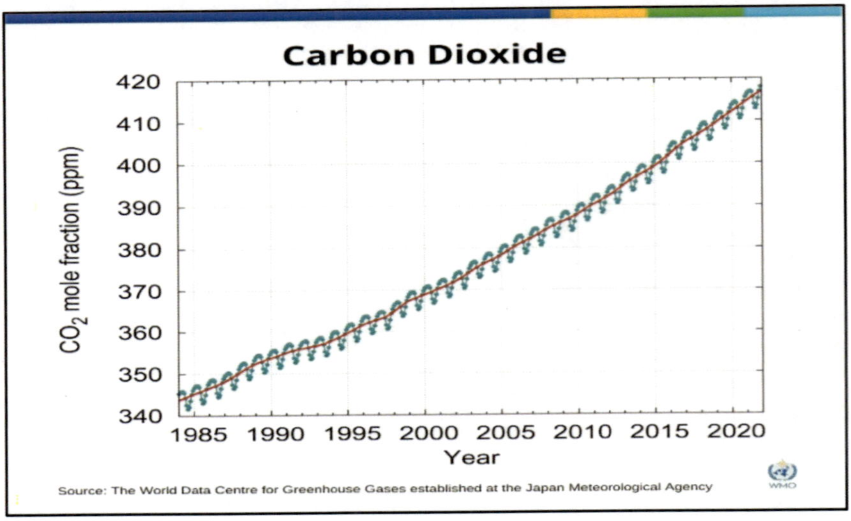

Fig. 3: Concentration of Greenhouse gases:(1985-2020)

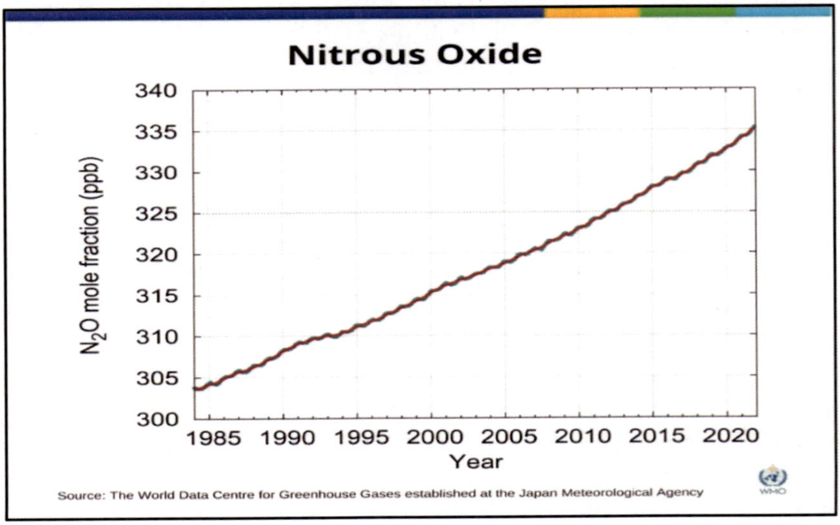

Fig. 4: Emission of Nitrous Oxide (1985-2020)

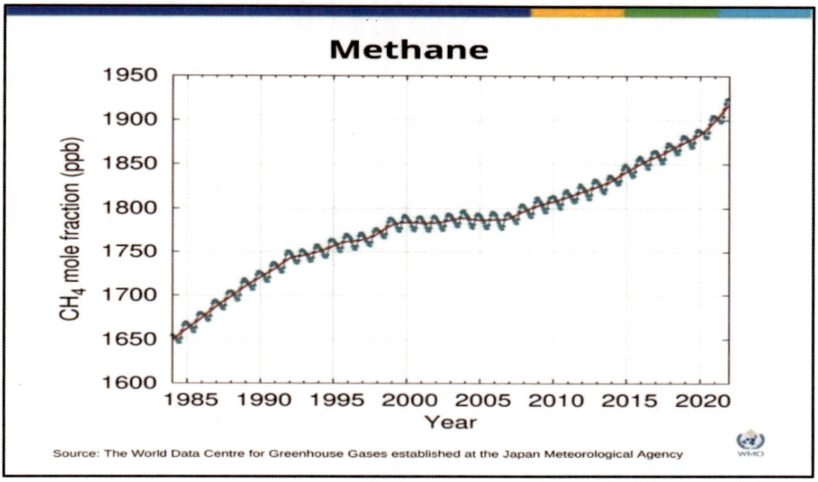

Fig. 5: Methane gas emission (1985-2020)

The increased concentrations of greenhouse gases in the atmosphere amplify the greenhouse effect (BBC, 2019). As these gases accumulate, they trap more heat, resulting in a rise in the Earth's average surface temperature. This phenomenon is known as global warming. The consequences of global warming include melting ice caps and glaciers, rising sea levels, more frequent and intense heatwaves, altered precipitation patterns, and shifts in ecosystems and biodiversity (BBC, 2023;NASA, 2019).

Understanding the greenhouse effect is essential for comprehending the causes and consequences of climate change (NASA, 2019). By recognizing the role of human activities in enhancing the greenhouse effect, we can take action to mitigate the emission of greenhouse gases and reduce our impact on the climate system. Transitioning to renewable energy sources, improving energy efficiency, adopting sustainable agricultural practices, and protecting and restoring natural carbon sinks such as forests are among the strategies to curb greenhouse gas emissions (Baxi, 2023; Brienen, et al., 2020).

2.2.3 Impacts of Climate Change

Climate change has wide-ranging and profound impacts on various aspects of the environment, ecosystems, human societies, and the economy (IPCC, 2022). These impacts are already being observed and are projected to escalate in severity and frequency if effective measures are not taken to mitigate and adapt to climate change. Some of the key impacts of climate change include:

i. *Rising temperatures*: Global warming leads to an increase in average temperatures worldwide. This rise in temperatures has direct consequences such as heatwaves, especially in urban areas, which pose risks to human health, especially vulnerable populations such as the elderly and children.

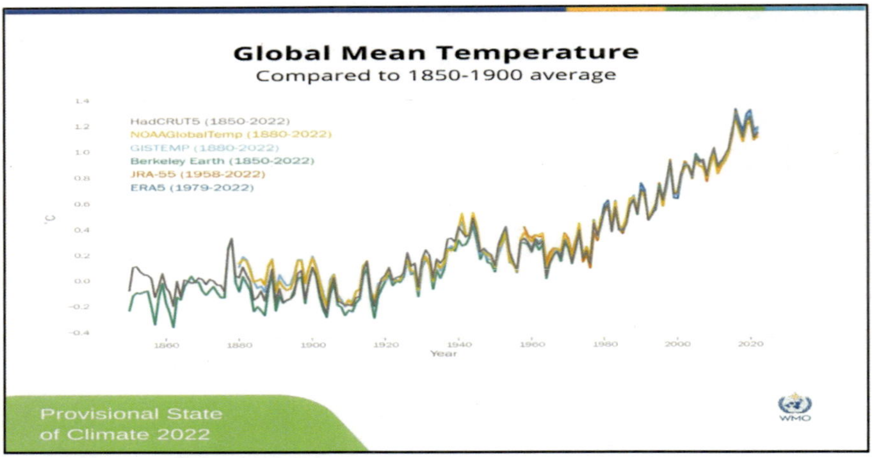

Fig. 6: Trend of change in Global Mean Temperature

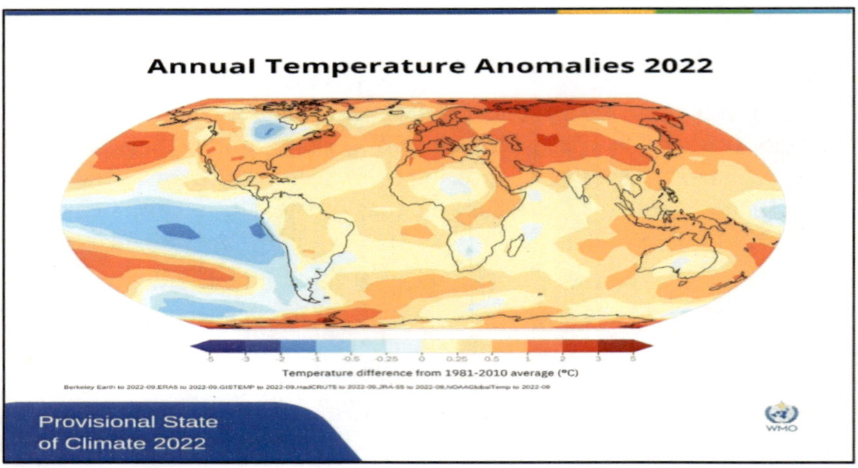

Fig. 7: Annual Temperature Anomalies

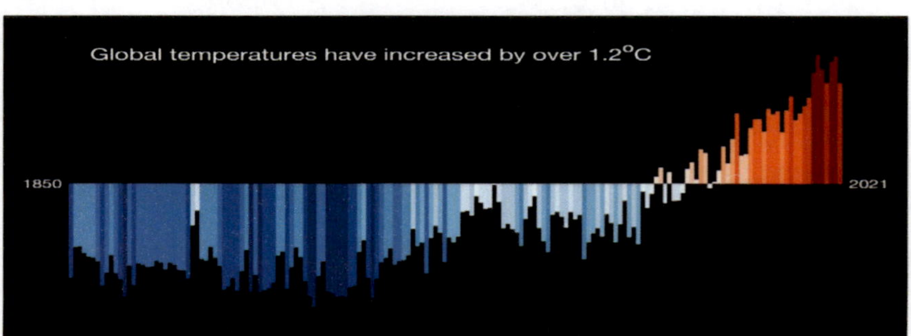

Fig. 8: Trend of increase in Global temperature

ii. *Melting ice caps and glaciers*: Higher temperatures cause the melting of polar ice caps and glaciers, leading to rising sea levels. This poses significant threats to coastal areas, increasing the risk of flooding, erosion, and the loss of valuable habitats.

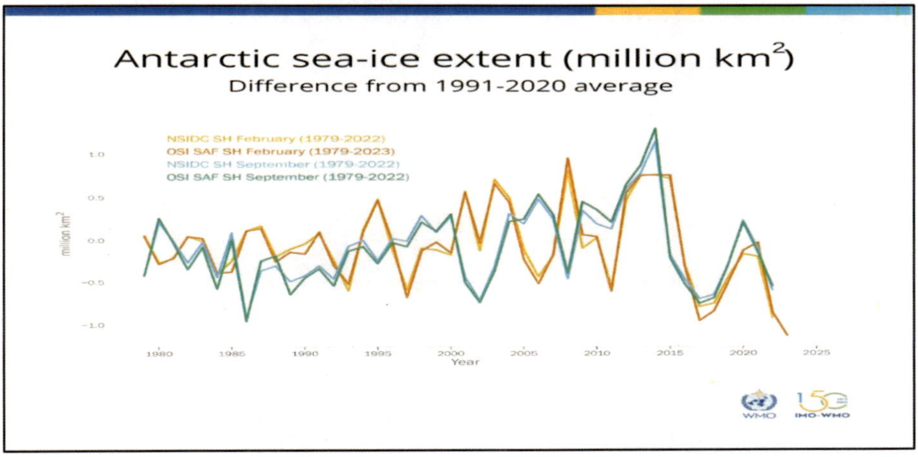

Fig. 9: Melting ice caps and glaciers

iii. *Sea-level rise*: As ice on land melts and the warming of seawater causes thermal expansion, sea levels are rising (NASA, 2018). Rising sea levels intensify the impacts of coastal flooding, particularly during storm events, and threaten low-lying islands and coastal communities. It also contributes to saltwater intrusion into freshwater resources, affecting drinking water supplies and agricultural productivity (Ki'Amber Thompson, 2019).

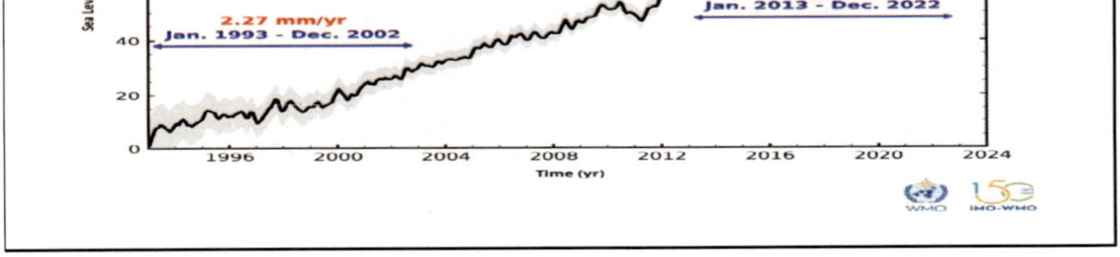

Fig. 10: Mean sea level is on rise

iv. *Extreme weather events*: Climate change increases the occurrence and severity of extreme weather phenomena like hurricanes, floods, droughts and heatwaves.These events have devastating consequences, leading to loss of life, displacement of communities, destruction of infrastructure, and significant economic costs.

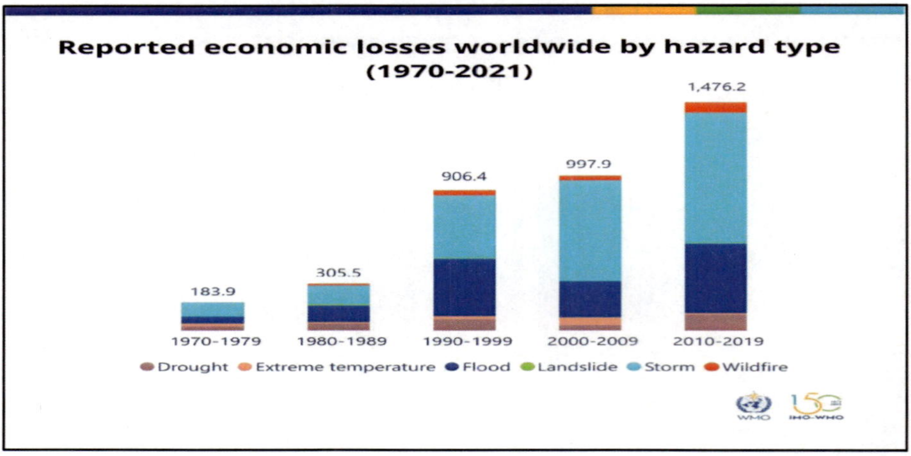

Fig. 11: Economic losses by Hazard types

v. *Shifts in ecosystems and biodiversity loss*: Climate change disrupts ecosystems and alters the distribution and behaviour of plant and animal species. Changes in temperature, rainfall patterns, and availability of resources impact habitats, migration patterns, and food chains. This can result in species extinctions, reduced biodiversity, and ecological imbalances.

vi. *Food security and agricultural impacts*: Changing climatic conditions affect agricultural productivity and food security (Squires and Gaur, 2020). Shifts in temperature and rainfall patterns can disrupt crop growth cycles, reduce yields, and make certain regions less suitable for traditional agricultural practices. Changes in precipitation patterns also affect water availability for irrigation, leading to water stress in agricultural regions (PNAS, 2016).

vii. *Public health risks*: Climate change impacts human health directly and indirectly. Heatwaves, extreme weather events, and changes in disease patterns affect the spread of infectious diseases, respiratory illnesses, and heat-related illnesses. Additionally, climate change can lead to food and water insecurity, malnutrition, and mental health challenges.

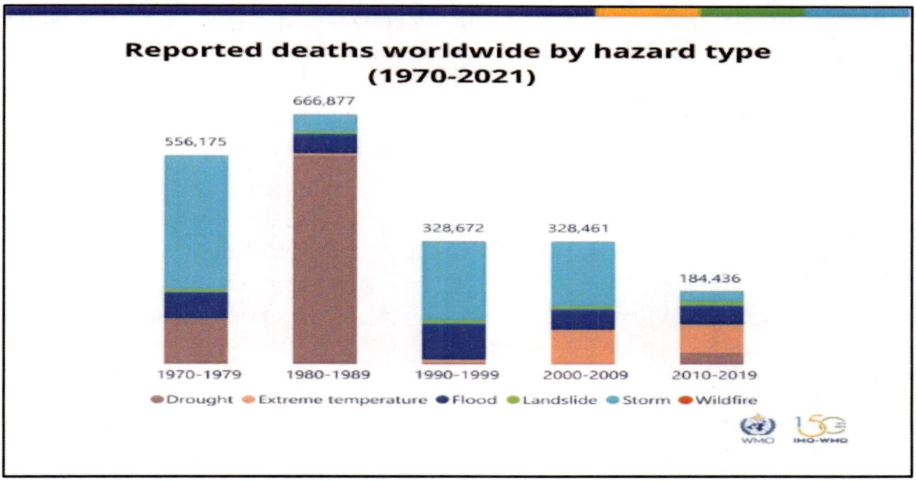

Fig. 12: Reported Deaths by Hazards

viii. *Social and economic disparities*: Climate change exacerbates existing social and economic inequalities. Vulnerable communities, including those in low-income regions, coastal areas, and developing countries, often have limited resources and infrastructure to adapt to climate impacts. This leads to disproportionate impacts on marginalized communities, deepening social disparities and creating climate justice concerns.

Addressing the impacts of climate change requires a comprehensive and multi-faceted approach. Mitigation efforts, such as reducing greenhouse gas emissions, transitioning to renewable energy sources, and promoting sustainable land use practices, are essential to limit the extent of climate change (IPCC, 2011; OECD, 2016). Adaptation measures, including improving infrastructure resilience, enhancing disaster preparedness, and implementing sustainable agricultural practices, are crucial to minimize the impacts and build resilience in vulnerable communities (WEF, 2021).

2.2.4 Mitigation Strategies

Mitigation strategies are efforts aimed at reducing greenhouse gas emissions and minimizing the causes of climate change. By implementing these strategies, we can work towards slowing down the rate of global warming and mitigating its impacts. Here are some key mitigation strategies:

i. *Transitioning to renewable energy sources*: It is essential to transit away from fossil fuels and increase the utilization of renewable energy sources like solar, hydroelectric, wind and geothermal power (Lattner, 2021; C-DAC, 2020). This transition reduces greenhouse gas emissions associated with electricity generation and promotes a sustainable energy system (Foster and Elzinga, 2015).

ii. *Improving energy efficiency*: Enhancing energy efficiency in buildings, transportation, and industrial processes helps reduce energy consumption and lowers greenhouse gas emissions (Wikipedia Contributors, 2018). This can be achieved through energy-efficient building designs, energy-saving appliances and equipment, and efficient transportation systems.

iii. *Promoting sustainable transportation*: Encouraging the use of public transportation, cycling, walking, and electric vehicles helps reduce emissions from the transportation sector (African Circular Business Alliance, 2023). Investing in well-connected public transit systems, promoting carpooling, and improving infrastructure for non-motorized transportation are important steps towards sustainable transportation.

iv. *Adopting sustainable land use practices*: Deforestation and land degradation contribute significantly to greenhouse gas emissions (Global Climate Action, 2020). Implementing sustainable land management practices, such as reforestation, afforestation, and sustainable agriculture, can help sequester carbon dioxide and mitigate emissions from land-use changes (Arneth et al., 2017).

v. *Enhancing waste management*: Implementing effective waste management practices, including recycling, composting, and waste-to-energy systems, reduces methane emissions from landfills. Proper waste management also conserves resources and reduces the need for raw materials, thereby reducing emissions from extraction and production processes.

vi. *Shifting to low-carbon industries*: Encouraging industries to adopt cleaner production technologies and processes can significantly reduce greenhouse gas emissions (C2ES, 2018). Promoting energy-efficient manufacturing, investing in research and development of low-carbon technologies, and implementing sustainable industrial practices are key aspects of this strategy (IBEF, 2021).

vii. *Promoting sustainable agriculture*: Sustainable agricultural practices, such as organic farming, agroforestry, and precision farming techniques, can help reduce emissions from agricultural activities (Pallavi, 2023). These practices also enhance soil health, improve water management, and promote biodiversity conservation.

viii. *Fostering international cooperation*: Climate change is a global challenge that requires collaborative efforts. International cooperation, through agreements such as the Paris Agreement, is crucial for setting emissions reduction targets, sharing knowledge and resources (Maizland, 2022), and supporting developing countries in their transition to low-carbon economies (CSF, 2021).

Mitigation strategies not only help reduce greenhouse gas emissions but also offer numerous co-benefits (Barrett, 2014). These include improved air quality, enhanced energy security, job creation in the renewable energy sector, and opportunities for sustainable economic development.

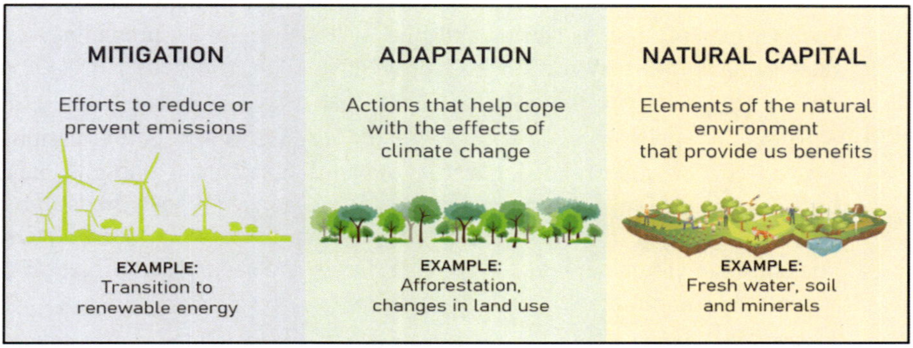

Fig. 13: Mitigation, Adaptation and Natural Capital

Fig. 14: Climate Change Adaptation and Resilience

2.2.5 Adaptation and Resilience

Adaptation measures are strategies and actions taken to adjust and respond to the impacts of climate change (Mehryar, 2022). While mitigation focuses on reducing greenhouse gas emissions, adaptation recognizes the need to prepare for and cope with the changes that are already occurring (IPCC, 2019) or are expected to occur in the future. Here are some key aspects of adaptation measures:

i. *Assessing climate risks*: Conducting comprehensive assessments of climate risks is essential to understand the potential impacts of climate change on various sectors and regions (NOAA, 2021). This involves analysing climate data, projecting future climate scenarios, and identifying vulnerabilities and hotspots that are most at risk.

ii. *Building climate resilience*: Enhancing the resilience of communities, ecosystems, and infrastructure is a crucial aspect of adaptation. This includes designing and retrofitting buildings and infrastructure to withstand extreme weather events, such as storms and floods. It also involves protecting and restoring natural ecosystems, which provide valuable services such as flood regulation and carbon sequestration.

iii. *Developing early warning systems*: Establishing robust early warning systems for weather-related hazards, such as hurricanes, droughts, and heatwaves, helps in timely preparedness and response (WMO, 2020). This includes investing in meteorological monitoring, forecasting technologies, and communication networks to provide timely and accurate information to vulnerable communities.

iv. *Improving water resource management*: Climate change affects the availability, quality, and distribution of water resources. Implementing adaptive water management strategies, such as water conservation measures, efficient irrigation techniques, and integrated water resource planning, helps ensure water security and resilience in the face of changing precipitation patterns and increased water stress (Leong, 2016).

v. *Enhancing agricultural practices*: Adapting agricultural systems to changing climatic conditions is crucial for ensuring food security and livelihoods. This includes promoting climate-smart agricultural practices like crop diversification, enhanced irrigation techniques, agroforestry, and the adoption of resilient crop varieties that can withstand climate challenges (IAEA, 2016). It also involves providing farmers with access to weather information, insurance, and financial support to cope with climate-related risks.

vi. *Strengthening coastal protection*: Coastal areas are particularly vulnerable to sea-level rise, storms, and erosion (US EPA, OW, 2015). Implementing measures such as beach nourishment, building sea walls, restoring coastal wetlands, and adopting sustainable coastal zone management practices help protect coastal communities, ecosystems, and infrastructure from climate impacts.

vii. *Enhancing public health preparedness*: Climate change can have significant impacts on public health, including the spread of diseases, heat-related illnesses, and mental health challenges. Strengthening public health systems, improving disease surveillance and response, and developing heatwave management plans help protect communities from climate-related health risks.

viii. *Promoting community engagement and social equity*: Ensuring that adaptation measures are inclusive and address the needs of vulnerable communities is crucial (DISD/DPIDG, 2021). Engaging local communities, indigenous peoples, and marginalized groups in decision-making processes, providing access to information and resources, and supporting capacity building initiatives contribute to more equitable and effective adaptation efforts.

Adaptation measures should be integrated into development planning processes, policies, and strategies at all levels, from local to national and international (GIZ, 2019). Collaboration among various stakeholders, including governments, communities, NGOs, and private sector entities, is essential for successful adaptation planning and implementation.

2.2.6 Role of Individuals and Communities

Individuals and communities play a crucial role in addressing climate change and global warming. While collective action and policy changes are essential, the actions and choices of individuals can significantly contribute to mitigating greenhouse gas emissions, adapting to climate impacts, and fostering a sustainable future. Here are key aspects of the role of individuals and communities:

i. *Energy conservation*: Individuals can reduce their carbon footprint by conserving energy in their homes and workplaces (Hahn, 2020). This includes using energy-efficient appliances and lighting, properly insulating buildings, and turning off lights and electronics when not in use (Afox, 2019). Conserving energy reduces greenhouse gas emissions associated with electricity generation (Shen et al., 2014).

ii. *Sustainable transportation*: Opting for sustainable transportation options, such as using public transit, carpooling, biking, or walking, helps reduce emissions from the transportation sector. By minimizing the use of private vehicles, individuals can contribute to improved air quality and reduced congestion while lowering their carbon footprint (WION, 2018).

iii. *Responsible consumption*: Making conscious choices about what we consume and how we consume it is crucial (Koch, 2022). Individuals can reduce their environmental impact by opting for sustainable and ethically produced goods, minimizing waste, and practicing the principles of reduce, reuse, and recycle. This includes avoiding single-use plastics, supporting local and organic products, and embracing a circular economy mind-set.

iv. *Plant-based diets*: The food system accounts for a significant portion of global greenhouse gas emissions (United Nations, 2022a). Choosing plant-based or vegetarian diets, or reducing meat consumption, can help lower emissions associated with livestock production and deforestation for agriculture. Individuals can opt for sustainable and locally sourced food options to reduce the carbon footprint of their diets (American Chemical Society, 2021).

v. *Advocacy and awareness*: Individuals can raise awareness and advocate for climate action in their communities. This can involve participating in local initiatives, joining environmental organizations, engaging in discussions with friends and family, and using social media platforms to share information about climate change and its impacts. By becoming climate ambassadors, individuals can inspire others to take action and influence policymakers to prioritize climate-related policies (European Climate Pact, 2022).

vi. *Community resilience and adaptation*: Building resilience at the community level is crucial for coping with climate impacts (UCDAVIS). Individuals can participate in community-based adaptation initiatives, such as planting trees, restoring natural habitats, supporting local climate resilience projects, and engaging in disaster preparedness and response efforts. By working together, communities can enhance their ability to adapt and thrive in a changing climate (ILO).

vii. *Knowledge sharing and capacity building*: Individuals can contribute to climate action by sharing knowledge and building capacity within their communities. This can involve organizing workshops, training programs, and educational events on climate change and sustainability (United Nations, 2021). By empowering others with information and skills, individuals can foster a culture of climate awareness and action (Brandvoice, 2021).

viii. *Voting and civic engagement*: Participating in elections and engaging in civic processes are crucial avenues for influencing climate policies. Individuals can vote for candidates who prioritize climate action, support climate-related initiatives and policies, and hold elected officials accountable for their climate commitments. By actively participating in democratic processes, individuals can drive change at local, regional, and national levels.

ix. *Collaboration and networking*: Engaging with like-minded individuals and organizations is essential for collective action on climate change. Individuals can join or form climate action groups, participate in community projects, and collaborate with local businesses, schools, and government agencies. By working together, individuals can leverage their collective power and amplify their impact.

It is important to recognize that individual actions, while significant, are not a substitute for systemic changes and policy interventions (Canzoneri, 2021). Individuals should also advocate for and support policy measures that address the root causes of climate change and create an enabling environment for sustainable practices (National Security Council, 2021).

2.2.7 The Importance of Scientific Research

Scientific research plays a crucial role in understanding and addressing climate change and global warming (C. Alfred Jacobson, 1916). It provides valuable insights into the causes, impacts, and potential solutions of climate change, informing policy decisions, guiding mitigation and adaptation strategies, and driving innovation. Here are key aspects highlighting the importance of scientific research in relation to climate change:

i. *Understanding the causes*: Scientific research helps us understand the underlying causes of climate change (Hazelton, 2021). Through rigorous observation, data collection, and analysis, scientists have identified the role of greenhouse gas emissions from human activities like burning non-renewable resources and deforestation, in driving global warming. This understanding is fundamental for designing effective mitigation measures and identifying areas where emissions reductions can have the greatest impact.

ii. *Assessing climate impacts*: Scientific research enables the assessment of climate change impacts on various sectors and ecosystems. This includes studying changes in temperature, precipitation patterns, sea-level rise, extreme weather events, biodiversity loss, and the implications for human health, agriculture, water resources, and vulnerable communities. By quantifying and projecting the impacts, researchers help policymakers and communities prepare for and adapt to changing conditions.

iii. *Modelling and projections*: Climate models are essential tools for understanding the complex interactions of the climate system and making projections about future climate scenarios (*Modelling and Projections*, 2021). Scientific research continuously improves these models, enhancing their accuracy and reliability. Climate models provide policymakers with critical information for decision-making, enabling them to assess the potential outcomes of different emission pathways and policy interventions.

iv. *Identifying mitigation strategies*: Scientific research informs the development and evaluation of mitigation strategies to reduce greenhouse gas emissions. It assesses the effectiveness and feasibility of different approaches, such as transitioning to renewable energy sources, improving energy efficiency, implementing carbon capture and storage technologies, and adopting sustainable land-use practices (IEA, 2021). Research helps identify opportunities, barriers, and trade-offs associated with different mitigation options (Sharifi, 2020).

v. *Supporting policy development*: Scientific research provides evidence-based information that supports the development of climate policies and agreements at local, national, and international levels (UN Secretary General, 2015). Policymakers rely on scientific assessments and recommendations to set emission reduction targets, design regulatory frameworks, and implement measures to incentivize sustainable practices (Lewandowski and Ullrich, 2023). Research helps policymakers make informed decisions and understand the potential consequences of different policy choices.

vi. *Innovation and technology development*: Scientific research drives innovation and technological advancements needed to address climate change. Researchers explore new materials, technologies, and processes that can reduce emissions, improve energy efficiency, and enhance resilience. This includes the development of renewable energy technologies, energy storage systems, climate-smart agriculture practices, and sustainable transportation solutions. Research also fosters collaboration between academia, industry, and government in finding sustainable solutions.

vii. *Monitoring and verification*: Scientific research plays a vital role in monitoring and verifying the progress of climate actions and agreements. Through systematic monitoring of greenhouse gas emissions, atmospheric composition, and climate indicators, researchers provide valuable data to assess the effectiveness of mitigation efforts and compliance with international agreements, such as the Paris Agreement (Wikipedia, 2023). Monitoring systems ensure transparency and accountability in climate action.

viii. *Bridging knowledge gaps*: Climate change is a complex and multidisciplinary issue that requires continuous research to address knowledge gaps and uncertainties. Scientific research helps fill these gaps by exploring new areas of study, refining methodologies, and integrating knowledge from various disciplines. This ongoing research contributes to a deeper understanding of climate change dynamics, improving the accuracy of projections and informing adaptive strategies.

ix. *Informing public discourse*: Scientific research contributes to informed public discourse and fosters a better understanding of climate change among the general public (Bruin and Morgan, 2019). Researchers communicate their findings through peer-reviewed publications, conferences, media engagements, and public outreach initiatives. By sharing their work and engaging in dialogue with stakeholders, scientists bridge the gap between scientific knowledge and public understanding, promoting evidence-based discussions on climate change.

Climate change and global warming pose significant challenges to our planet and require urgent and concerted action (Ahammed, 2021). The points discussed in this theoretical orientation emphasize the key aspects related to climate change and global warming, providing a comprehensive understanding of the issue.

2.3 Uncertainty

Uncertainty is a fundamental aspect of life, and its understanding is crucial in various fields, including agriculture. Uncertainty refers to the lack of complete knowledge or predictability about a particular event or outcome. It encompasses a range of factors that can affect decision-making processes and outcomes. By recognizing and managing uncertainty, individuals and organizations can navigate challenges and make more informed choices.

2.3.1 Understanding Uncertainty

Uncertainty can be defined as the state of not knowing with certainty what the future holds. It arises from incomplete information, complex systems, and the inherent variability of natural and social processes. Uncertainty is not synonymous with risk, which typically involves quantifiable probabilities and potential losses or gains. Instead, uncertainty deals with the unknown and the unpredictable.

2.3.2 Types of Uncertainty

There are several types of uncertainty that can impact agricultural decision making. Epistemic uncertainty arises from a lack of knowledge or understanding about a particular phenomenon. It can be reduced through research, data collection, and improved scientific understanding. Aleatory uncertainty, on the other hand, stems from inherent variability and randomness in natural processes, such as weather patterns or pest outbreaks. This type of uncertainty is irreducible and requires strategies to adapt and cope with its effects.

2.3.3 Importance of Managing Uncertainty

Managing uncertainty is crucial for agricultural decision making as it helps farmers and policymakers make informed choices and mitigate potential risks. By understanding the sources and types of uncertainty, stakeholders can develop strategies to minimize negative impacts and maximize opportunities. Effective management of uncertainty can lead to increased productivity, improved resilience, and sustainable agricultural practices.

2.3.4 Chaos Theory and the Butterfly Effect

Chaos theory is a branch of mathematics that deals with complex, nonlinear systems that are highly sensitive to initial conditions. It explores the behaviour of systems that appear random and unpredictable but are governed by underlying deterministic processes. One of the key concepts in chaos theory is the butterfly effect.

2.3.4.1 Overview of Chaos Theory

Chaos theory suggests that small changes in initial conditions can lead to significant differences in outcomes over time. Complex systems, such as weather patterns, ecosystems, or financial markets, can exhibit chaotic behaviour. While the underlying processes may be deterministic, the sensitivity to initial conditions makes long-term predictions challenging.

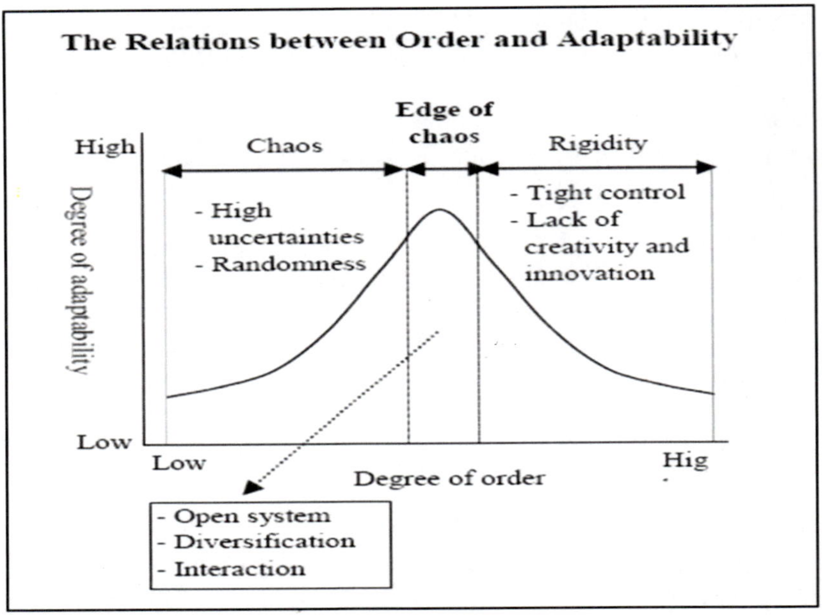

Fig. 15: Relation between Order and Adaptability

2.3.4.2 The Butterfly Effect and its Significance

The butterfly effect is a metaphor used to illustrate the sensitivity to initial conditions in chaotic systems. It suggests that a butterfly flapping its wings in one part of the world can potentially influence the course of a hurricane in another part of the world. This concept highlights the interconnectedness and nonlinearity of complex systems, where small perturbations can have cascading effects.

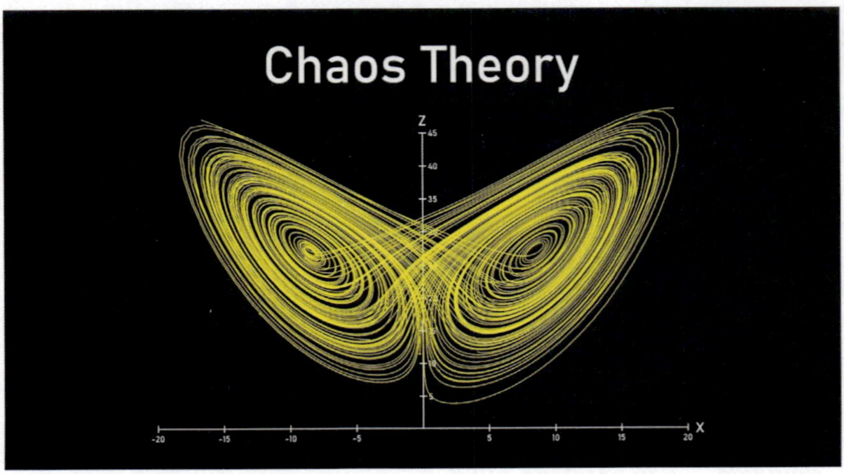

Fig. 16: Chaos Theory

2.3.5 Climate Uncertainty and Agricultural Decision Making

Climate change poses significant uncertainties for agricultural decision making. It refers to long-term shifts in weather patterns, including changes in temperature, precipitation, and extreme events. These changes can have profound impacts on agricultural productivity, water availability, and pest dynamics.

2.3.5.1 Impact of Climate Change on Agriculture

Climate change can disrupt crop yields, water availability, and the prevalence of pests and diseases. Farmers face difficulties due to the rising temperatures, changing rainfall patterns, and a higher occurrence of extreme weather events. Heat stress, droughts, floods, and altered pest populations can reduce agricultural productivity and increase production risks.

2.3.5.2 Challenges Faced by Farmers

Farmers face numerous challenges in adapting to climate change. They must contend with uncertain growing seasons, increased pest pressure, and changing water availability. Small-scale farmers, in particular, often lack access to resources, technologies, and financial support needed to adapt to climate uncertainties. These challenges can lead to reduced incomes, food insecurity, and vulnerability for farming communities.

2.3.5.3 Strategies for Managing Climate Uncertainty

To manage climate uncertainty, farmers can employ various strategies. These may include diversifying crops, adopting climate-resilient varieties, improving water management practices, and implementing precision agriculture techniques. Additionally, integrating climate information and forecasts into decision-making processes can help farmers anticipate and adapt to climate variability. Collaborative efforts between farmers, researchers, and policymakers are essential to developing and implementing effective strategies for managing climate uncertainty in agriculture.

2.3.6 Market Uncertainty

Market uncertainty refers to the unpredictability and volatility of agricultural markets. Factors such as fluctuating commodity prices, supply and demand dynamics, trade policies, and global economic conditions contribute to market uncertainty in agriculture.

2.3.6.1 Volatility in Agricultural Markets

Agricultural markets can experience significant price fluctuations due to various factors. These include weather events affecting crop yields, changes in consumer preferences, geopolitical tensions impacting trade, and shifts in currency exchange rates. These fluctuations can create uncertainty for farmers in terms of income generation and financial planning.

2.3.6.2 Factors Affecting Market Uncertainty

Market uncertainty in agriculture can be influenced by several factors. These include weather-related events affecting crop production, political decisions impacting trade agreements and tariffs, changes in consumer demands and preferences, and macroeconomic conditions such as inflation and interest rates. Additionally, market uncertainty can be influenced by technological advancements, market regulations, and the actions of market participants.

2.3.6.3 Risk Management Techniques

To cope with market uncertainty, farmers can employ risk management techniques. These include forward contracting, hedging, crop insurance, diversification of crops and markets, and building strong relationships with buyers and suppliers. By implementing these strategies, farmers can mitigate potential losses and stabilize their income in the face of market volatility.

2.3.7 Economic Uncertainty

Economic uncertainty refers to the unpredictability and instability of economic conditions. In the context of agriculture, it encompasses factors such as fluctuating input costs, changes in government policies, and macroeconomic fluctuations that affect agricultural production and profitability.

2.3.7.1 Economic Factors Influencing Agriculture

Agricultural production and profitability are influenced by various economic factors. These include input costs such as fuel, fertilizer, and labour, as well as commodity prices, interest rates, exchange rates, inflation, and taxation policies. Economic conditions can affect farmers' decisions regarding resource allocation, investment in technology, and market participation.

2.3.7.2 Macroeconomic Uncertainty and Agriculture

Macroeconomic uncertainty, such as recessions, currency devaluations, or global financial crises, can have far-reaching effects on agriculture. Changes in consumer spending patterns, reduced access to credit, and shifts in government support programs can impact farmers' incomes, market demand, and investment opportunities.

2.3.7.3 Coping with Economic Uncertainty

To cope with economic uncertainty, farmers can adopt strategies such as cost management, diversification of income sources, improving efficiency and productivity, and seeking financial assistance through government programs or agricultural cooperatives. Additionally, staying informed about economic trends and engaging in market intelligence can help farmers make informed decisions in uncertain economic conditions.

2.3.8 Uncertainty and Resilience

Uncertainty poses challenges to agricultural systems, but it also presents opportunities to build resilience. Resilience refers to the ability of a system or community to absorb shocks, adapt to changes, and maintain essential functions and productivity in the face of disturbances.

2.3.8.1 Building Resilience in Agriculture

Building resilience in agriculture involves integrating adaptive strategies, diversifying resources, fostering knowledge exchange and learning, and enhancing social networks and collaboration. This can include implementing sustainable farming practices, investing in infrastructure for water management, adopting climate-smart technologies, and promoting biodiversity on farms.

2.3.8.2 Adaptive Strategies for Uncertain Times

Adaptive strategies for uncertain times involve continually monitoring and assessing the changing conditions, being flexible and responsive to new information, and implementing timely adjustments to management practices. These strategies may include adjusting planting schedules based on weather forecasts, diversifying crop varieties, practicing soil conservation techniques, and implementing precision agriculture technologies.

2.3.9 Farmers' Suicide: Understanding the Issue

Farmers' suicide is a tragic phenomenon that has gained attention in recent years. It refers to the occurrence of suicides among farmers, often attributed to various factors such as financial distress, crop failure, debt burden, and psychological stress.

2.3.9.1 Factors Contributing to Farmers' Suicide

Multiple factors contribute to farmers' suicide. Economic factors, including low agricultural income, increasing debt levels, lack of access to credit, and market volatility, can create financial stress and desperation among farmers. Environmental factors, such as droughts, floods, and pest outbreaks, can lead to crop failure and loss of livelihood. Social factors, including isolation, lack of support systems, and stigma surrounding mental health, can exacerbate mental health issues among farmers.

2.3.9.2 Addressing Mental Health and Well-being

Addressing mental health and well-being is crucial in combating farmers' suicide. It involves raising awareness about mental health issues, reducing stigma, and improving access to mental health services in rural areas. Providing counselling services, support groups, and training programs to enhance coping skills can help farmers manage stress and emotional challenges effectively.

2.3.10 Stress in Agriculture: Causes of Stress Among Farmers

Farmers face numerous stressors in their profession. Uncertainty related to weather conditions, market fluctuations, and financial stability can create chronic stress. Physical demands, long working hours, and limited time for leisure activities can contribute to stress. Additionally, social pressures, such as the expectation to carry on the family farming tradition or meet community expectations, can add to the stress burden.

2.3.10.1 Impacts of Stress on Agricultural Communities

Stress in agricultural communities can have detrimental effects on individuals, families, and the overall community. It can lead to mental health disorders, strained relationships, substance abuse, and reduced quality of life. Stress can also affect decision-making abilities, productivity, and innovation in agriculture, potentially perpetuating a cycle of challenges and hardships.

2.3.10.2 Promoting Mental Wellness in Agriculture

Promoting mental wellness in agriculture involves creating supportive environments, providing resources for stress management and mental health support, and fostering a culture of open communication. Developing programs that address mental health needs, providing training on stress management and coping strategies, and establishing peer support networks can contribute to the well-being of farmers and agricultural communities.

2.3.11 Disillusionment and Off-Farm Migration: Reasons and Consequences

Disillusionment among farmers refers to feelings of dissatisfaction, frustration, or loss of hope regarding the agricultural profession. Various factors contribute to disillusionment, including financial challenges, unpredictable market conditions, diminishing profitability, and limited opportunities for growth and advancement. Environmental factors, such as climate change impacts and resource depletion, can also contribute to disillusionment.

Off-farm migration occurs when farmers and rural residents leave their agricultural livelihoods to seek employment or opportunities in non-agricultural sectors. It can lead to significant changes in rural communities, including depopulation, loss of traditional knowledge and skills, and shifts in local economies. Off-farm migration may alleviate some financial pressures for individuals, but it can also result in social and cultural disintegration within rural communities.

2.3.12 Supporting Farming Communities

Supporting farming communities involves implementing policies and programs that address the root causes of disillusionment and provide alternative avenues for income generation and career development. This can include promoting diversification of rural economies, improving access to education and training programs, and facilitating the adoption of sustainable and innovative agricultural practices. Additionally, fostering social networks, community cohesion, and preserving cultural heritage can contribute to the overall well-being and resilience of farming communities.

2.4 Agricultural Dynamics

Agricultural dynamics play a crucial role in shaping the agricultural sector. The term "agricultural dynamics" refers to the changes, shifts, and fluctuations that occur within the agricultural industry over time. These dynamics encompass a wide range of factors and mechanisms that influence the agricultural system, including technological advancements, environmental factors, market forces, policies, socioeconomic factors, resilience, knowledge systems, and interdisciplinary approaches.

Fig. 17: Change in Agricultural landscaping is prominent

The purpose of this theoretical orientation is to provide a framework for understanding and analysing agricultural dynamics in a comprehensive manner. By examining and studying the underlying factors and mechanisms driving agricultural dynamics, researchers and policymakers can gain insights into the dynamics of change within the agricultural sector and make informed decisions to foster sustainable agricultural development.

The study of agricultural dynamics is important due to the significant role that agriculture plays in global food production, rural livelihoods, and overall economic development. Understanding the dynamics within the agricultural sector is essential for ensuring food security, enhancing productivity and efficiency, addressing environmental challenges, and promoting equitable and sustainable agricultural practices.

Fig. 18: Study on uncertainty and agricultural dynamics

By examining the various factors and mechanisms that shape agricultural dynamics, this theoretical orientation aims to shed light on the complexity and interconnectedness of the agricultural system. It recognizes that agriculture is not a standalone entity but rather a complex system influenced by multiple factors and stakeholders. Therefore, adopting a systemic approach is crucial to comprehending the dynamics within the agricultural sector.

In the following sections, we will delve into the key factors and mechanisms that contribute to agricultural dynamics. We will explore the role of technological innovations, environmental factors, market forces, policies, socioeconomic factors, resilience, knowledge systems, and interdisciplinary approaches in shaping agricultural dynamics. By examining each of these components, we can develop a holistic understanding of the dynamics within the agricultural sector and their implications for sustainable agricultural development.

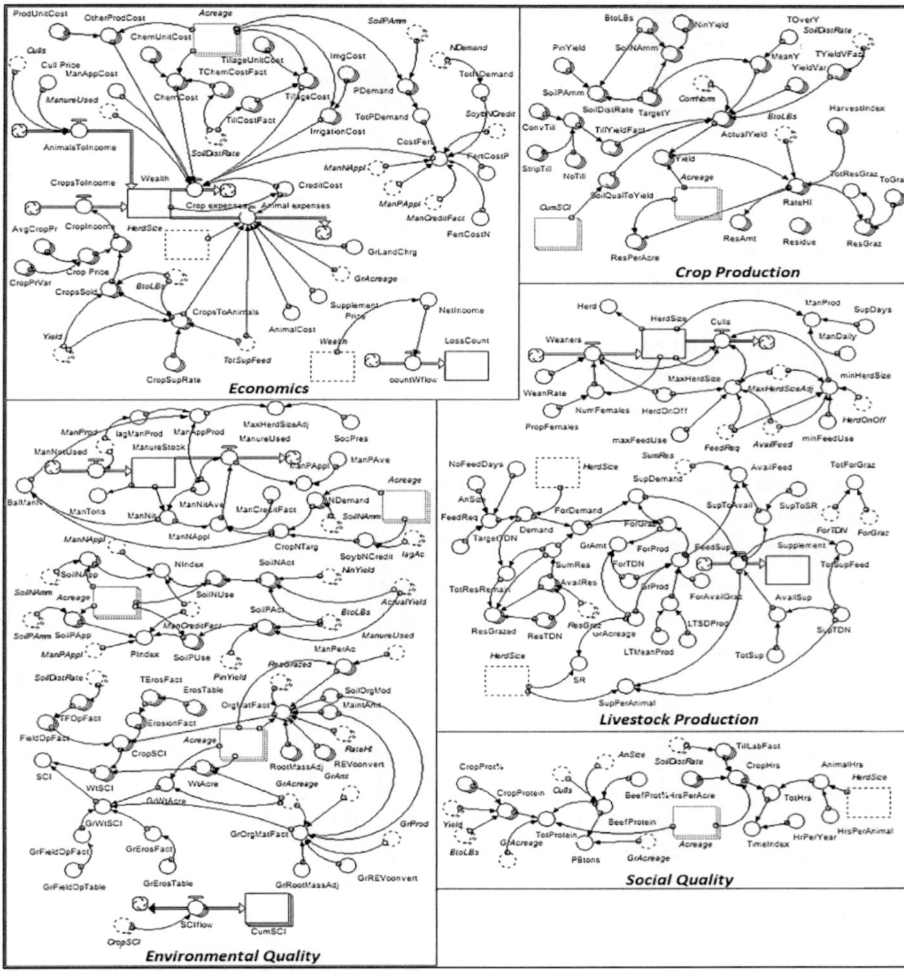

Fig. 19: Network Analysis : Environment and Sustainable Development

2.4.1 Systemic Approach

Adopting a systemic approach is essential for understanding the dynamics within the agricultural sector. It involves recognizing and analysing the complex interactions, interdependencies, and feedback loops that exist among the various components of the agricultural system. By examining the agricultural system as a whole, rather than focusing solely on individual components, researchers and policymakers can gain a comprehensive understanding of how different factors and mechanisms influence agricultural dynamics.

The agricultural system comprises multiple interconnected components that collectively contribute to its functioning. These key components include:

i. *Farmers*: Farmers are at the core of the agricultural system. They engage in agricultural production, make decisions about crop selection, livestock management, and utilize farming techniques. The characteristics, knowledge, and practices of farmers significantly impact agricultural dynamics.

ii. *Agricultural Practices*: The methods and techniques employed in agricultural production are critical factors in agricultural dynamics. These practices encompass a range of activities such as land preparation, crop cultivation, irrigation, pest and disease management, and livestock rearing. Changes in agricultural practices, such as the adoption of sustainable farming methods or the introduction of precision agriculture, can lead to shifts in productivity, resource efficiency, and environmental sustainability.

iii. *Technology*: Technological advancements have a profound impact on agricultural dynamics. Innovations in agricultural machinery, equipment, tools, and biotechnology have revolutionized farming practices, increasing productivity, efficiency, and sustainability. Technologies such as precision agriculture, drones, remote sensing, and genetic engineering have the potential to transform agricultural systems and shape their dynamics.

iv. *Markets*: Agricultural dynamics are closely linked to market forces. Supply and demand dynamics, price fluctuations, and global trade patterns influence the choices made by farmers and the overall agricultural landscape. Changes in consumer preferences, market regulations, and trade policies can impact agricultural production, market access, and the profitability of farmers.

v. *Policies*: Policies and governance structures play a crucial role in shaping agricultural dynamics. Agricultural policies related to subsidies, trade regulations, land ownership, research and development funding, and environmental regulations have direct and indirect effects on agricultural practices and outcomes. Policy decisions can incentivize or hinder innovation, influence resource allocation, and impact the resilience of the agricultural system.

vi. *Environmental Factors*: Environmental factors exert a significant influence on agricultural dynamics. Climate change, soil degradation, water availability, and biodiversity loss affect agricultural productivity, resource availability, and risk management. Understanding the interactions between agriculture and the environment is vital for developing sustainable agricultural practices and mitigating potential risks.

2.4.2 Technological Innovations

Technological advancements have been instrumental in shaping agricultural dynamics and transforming the agricultural sector. These innovations encompass a wide range of developments, including the introduction of new agricultural machinery, precision farming techniques, biotechnology, and digital tools.

Technological innovations have had profound impacts on productivity, efficiency, and sustainability in agriculture.

i. *Agricultural Machinery*: The introduction of advanced agricultural machinery has revolutionized farming practices. Tractors, harvesters, seeders, and irrigation systems have increased the efficiency of labour-intensive tasks, reduced manual effort, and improved overall productivity. Mechanization has enabled farmers to cultivate larger areas of land, manage crops more effectively, and enhance their output.

ii. *Precision Farming*: Precision farming techniques leverage technology to optimize agricultural practices at the field level. Remote sensing, geographic information systems (GIS), and global positioning systems (GPS) are used to collect and analyse data about soil characteristics, crop growth, and environmental conditions. This information allows farmers to tailor their practices, such as variable rate application of fertilizers and pesticides, precise irrigation, and targeted crop management, leading to improved resource efficiency and reduced environmental impact.

iii. *Biotechnology*: Biotechnological innovations have had a significant impact on agricultural dynamics. Genetically modified organisms (GMOs) have been developed to enhance crop traits such as resistance to pests and diseases, tolerance to environmental stresses, and improved nutritional content. Biotechnology has the potential to increase crop yields, reduce chemical inputs, and address specific challenges faced by farmers, contributing to improved agricultural productivity and sustainability.

iv. *Digital Tools and Data Analytics*: The emergence of digital tools and data analytics has transformed agricultural practices and decision-making processes. Farmers can now access real-time weather data, market information, and crop monitoring tools through mobile applications and online platforms. Data analytics enable farmers to make data-driven decisions, optimize resource allocation, manage risks, and improve productivity. Moreover, digital platforms facilitate knowledge exchange, connecting farmers with experts and enabling the sharing of best practices.

The adoption of technological innovations in agriculture has several implications for agricultural dynamics:

i. *Productivity*: Technological advancements have led to increased agricultural productivity by enabling more efficient use of resources, reducing losses, and optimizing production practices. Higher productivity contributes to food security, economic growth, and rural development.

ii. *Efficiency*: Technology has enhanced resource efficiency in agriculture by optimizing the use of inputs such as water, fertilizers, and pesticides. Precision farming techniques, for instance, allow for targeted application of inputs based on specific field requirements, minimizing waste and environmental impacts.

iii. *Sustainability*: Technological innovations have the potential to improve the sustainability of agricultural practices. For example, precision farming techniques reduce the use of chemicals, minimize soil erosion, and promote conservation practices. Biotechnology can help develop crop varieties with improved resilience to climate change, reducing the need for chemical inputs and contributing to environmental sustainability.

iv. *Adaptation*: Technological innovations enable farmers to adapt to changing environmental conditions and market demands. They provide tools and knowledge to mitigate risks associated with climate variability, pests, and diseases, enhancing the resilience of agricultural systems.

However, the adoption of technological innovations in agriculture also presents challenges and considerations, such as access to technology, affordability, capacity building, and potential environmental and social impacts. It is crucial to ensure that technological advancements are accessible, equitable, and aligned with sustainable development goals.

2.4.3 Environmental Factors

Environmental factors play a critical role in shaping agricultural dynamics. The interaction between agriculture and the environment is complex and multifaceted, with environmental conditions and changes significantly influencing agricultural practices, productivity, and sustainability. Key environmental factors that impact agricultural dynamics include climate change, soil degradation, water scarcity, and biodiversity loss.

i. *Climate Change*: The agricultural systems worldwide face significant challenges due to climate change. The increasing temperatures, changing patterns of rainfall, more frequent occurrence of extreme weather events, and changes in pest and disease dynamics collectively impact the productivity and stability of agriculture. Changes in growing seasons and temperature regimes can impact crop yields and quality, disrupt pollination, and increase the vulnerability of crops to pests and diseases. Understanding the impacts of climate change on agriculture is crucial for adapting farming practices, developing resilient crop varieties, and implementing climate-smart strategies.

ii. *Soil Degradation*: Soil degradation, including erosion, nutrient depletion, and soil pollution, poses a threat to agricultural productivity and long-term sustainability. Unsustainable farming practices, such as intensive tillage, overuse of chemical inputs, and improper land management, contribute to soil degradation. Degraded soils have reduced water-holding capacity, lower fertility, and decreased nutrient availability, negatively impacting crop growth. Implementing soil conservation practices, such as conservation tillage, agroforestry, and organic farming, is vital for preserving soil health and ensuring sustainable agricultural production.

iii. *Water Scarcity*: Water scarcity is a significant constraint on agricultural production in many regions. As water resources become increasingly limited, agricultural systems face challenges in irrigation, livestock management, and crop selection. Sustainable water management practices, such as efficient irrigation methods, water recycling, and rainwater harvesting, are essential for maximizing water use efficiency and minimizing the environmental impact of agriculture. Additionally, developing drought-tolerant crop varieties and improving water storage and distribution infrastructure can help mitigate the effects of water scarcity.

iv. *Biodiversity Loss*: Agricultural activities have often led to the loss of biodiversity, including the destruction of natural habitats, the loss of native plant and animal species, and the decline of pollinators. Biodiversity loss can have adverse consequences for agricultural systems, as it affects ecosystem services such as pollination, pest control, and nutrient cycling. Conservation and restoration of biodiversity within agricultural landscapes, promoting agro-ecological approaches, and preserving natural habitats adjacent to farmland are important strategies for maintaining ecosystem resilience and supporting sustainable agricultural practices.

Understanding the interactions between agriculture and the environment is crucial for developing sustainable agricultural practices and mitigating potential risks. Sustainable agriculture aims to optimize productivity while minimizing negative environmental impacts. It involves adopting practices such as agro ecology, integrated pest management, water-efficient irrigation, and soil conservation techniques. By integrating environmental considerations into agricultural decision-making processes, policymakers and farmers can reduce the ecological footprint of agriculture, enhance resilience to environmental challenges, and contribute to long-term food security and sustainable development.

2.4.4 Market Factors

Market forces have a significant impact on agricultural dynamics, shaping the choices made by farmers and influencing the overall agricultural landscape. Understanding the interplay between supply and demand dynamics, price fluctuations, and global trade patterns is crucial for comprehending how market forces drive agricultural dynamics.

i. *Supply and Demand Dynamics*: The interaction between agricultural supply and demand is a fundamental driver of agricultural dynamics. Supply factors include agricultural production levels, technological advancements, input availability, and resource constraints. Demand factors encompass consumer preferences, population growth, dietary patterns, and income levels. Changes in supply and demand can affect market prices, crop selection, production decisions, and investment patterns within the agricultural sector.

ii. *Price Fluctuations*: Agricultural markets are subject to price fluctuations due to various factors, including weather conditions, changes in global commodity markets, currency exchange rates, and government policies. Price volatility can influence farmers' profitability, investment decisions, and production choices. High prices may incentivize increased production, while low prices may lead to reduced production or shifts to alternative crops. Price signals influence farmers' allocation of resources, adoption of new technologies, and engagement in market-oriented agricultural practices.

iii. *Global Trade Patterns*: International trade in agricultural commodities plays a significant role in agricultural dynamics. Global trade affects domestic agricultural markets by influencing prices, market access, and competition. Trade policies, tariffs, subsidies, and non-tariff barriers influence the flow of agricultural products across borders, impacting farmers' market opportunities and agricultural production decisions. Changes in global trade patterns can lead to shifts in crop production, specialization, and investment patterns in different regions.

iv. *Consumer Preferences*: Consumer preferences and demands for agricultural products, including quality, safety, sustainability, and ethical considerations, shape agricultural dynamics. Changing consumer preferences can drive shifts in production systems, product differentiation, and marketing strategies. Farmers and agribusinesses respond to consumer demands by adopting practices such as organic farming, sustainable production methods, and traceability systems.

Market forces influence agricultural dynamics through several mechanisms:

i. *Investment and Technology Adoption*: Market signals, including price incentives and expected profitability, influence farmers' decisions regarding investment in technologies, equipment, and farm infrastructure. Favourable market conditions can drive adoption of new technologies, such as precision farming systems, genetically modified crops, or improved irrigation methods, to enhance productivity and competitiveness.

ii. *Crop Selection and Production Decisions*: Market prices for different agricultural commodities influence farmers' crop selection decisions. When prices are favourable, farmers may shift towards crops that offer higher returns, while lower prices may lead to reduced production of certain crops. Market signals guide farmers in choosing which crops to grow and allocate their resources accordingly.

iii. *Market Access and Value Chains*: Market forces impact farmers' access to markets and value chains. Factors such as transportation infrastructure, logistics, market information, and market power of buyers influence farmers' ability to participate in markets and capture value-added opportunities. Access to competitive markets is crucial for smallholder farmers to enhance their incomes and improve their livelihoods.

iv. *Innovation and Entrepreneurship*: Market forces drive innovation and entrepreneurship in agriculture. Market opportunities and consumer demands incentivize farmers and agribusinesses to develop and adopt new technologies, products, and marketing strategies. Entrepreneurial activities, such as value-added processing, agri-tourism, or organic farming, can arise in response to market forces and shape agricultural dynamics.

Market forces have both positive and negative implications for agricultural dynamics. They can create opportunities for farmers to improve profitability, access new markets, and adopt innovative practices. However, they can also lead to challenges such as price volatility, market concentration, and inequitable market access. Ensuring fair competition, promoting transparent markets, providing market information, and supporting farmers' capacity to navigate market forces are important for creating a conducive environment for sustainable agricultural development.

2.4.5 Policy and Governance

Policies and governance structures play a crucial role in shaping agricultural dynamics at different levels, from national to regional and international. Agricultural policies encompass a range of measures and regulations that govern various aspects of agricultural production, trade, land ownership, research and development, and environmental management. Understanding the impact of policies and governance on agriculture is essential for comprehending the factors that shape agricultural dynamics.

i. *Subsidies and Support Programs*: Governments often implement agricultural subsidies and support programs to provide financial assistance and incentives to farmers. Subsidies can take various forms, such as direct payments, price supports, input subsidies, and crop insurance. These policies aim to stabilize farmers' incomes, promote agricultural production, and ensure food security. The design and implementation of subsidy programs can influence farmers' decision-making, resource allocation, and the adoption of specific agricultural practices.

ii. *Trade Regulations and Agreements*: Trade regulations and agreements have a significant impact on agricultural dynamics, particularly in the context of international trade. Tariffs, quotas, sanitary and phytosanitary measures, and technical regulations can affect market access, competitiveness, and export opportunities for agricultural products. Regional and international trade agreements, such as free trade agreements or regional integration initiatives, can shape the patterns of agricultural production, specialization, and trade flows.

iii. *Land Ownership and Tenure Systems*: Policies related to land ownership and tenure systems have implications for agricultural dynamics, particularly in terms of access to land, land use patterns, and agricultural investments. Land reform policies, land titling, land consolidation, and regulations governing land transactions can influence the structure of agricultural holdings, promote sustainable land use practices, and address issues of equity and social justice in rural areas.

iv. *Research and Development Funding*: Government investment in agricultural research and development (R&D) plays a crucial role in shaping agricultural dynamics. Funding for agricultural R&D supports the development and adoption of improved technologies, crop varieties, and management practices. Investments in agricultural R&D can enhance productivity, resource efficiency, and sustainability in agriculture, leading to advancements in agricultural systems and practices.

v. *Environmental Regulations and Conservation Policies*: Environmental regulations and conservation policies are instrumental in shaping agricultural dynamics. Policies aimed at mitigating environmental impacts, promoting sustainable land use, and conserving natural resources can influence agricultural practices and production systems. Regulations related to pesticide use, water quality, land conservation, and ecosystem services contribute to sustainable agricultural development and the preservation of natural resources.

Effective policy and governance frameworks in agriculture should consider the following:

i. *Coherence and Integration*: Agricultural policies should be coherent and integrated across different sectors and policy domains. Policies related to agriculture, trade, environment, rural development, and social welfare should be aligned to avoid conflicting objectives and promote sustainable agricultural systems.

ii. *Stakeholder Engagement*: Inclusive and participatory policymaking processes that involve stakeholders, including farmers, agribusinesses, civil society organizations, and researchers, are crucial for designing effective agricultural policies. Stakeholder engagement ensures that policies address the needs and aspirations of diverse agricultural actors and foster ownership and implementation.

iii. *Knowledge and Evidence-Based Decision-Making*: Policy formulation should be based on sound scientific knowledge, evidence, and rigorous analysis. Evidence-based decision-making helps ensure that policies are effective, efficient, and responsive to the complex challenges facing the agricultural sector.

iv. *Monitoring and Evaluation*: Regular monitoring and evaluation of policy implementation and impact are vital for assessing the effectiveness and efficiency of agricultural policies. Monitoring allows for timely adjustments and learning, while evaluation provides insights into the outcomes, unintended consequences, and effectiveness of policies.

v. *International Cooperation*: Given the interconnectedness of agricultural systems and global trade, international cooperation is crucial for addressing cross-border challenges and fostering sustainable agricultural development. Collaboration between countries, knowledge sharing, and policy coordination contribute to creating an enabling global environment for agricultural dynamics.

2.4.6 Socioeconomic Factors

Socioeconomic factors play a critical role in shaping agricultural dynamics, as they influence the social and economic context within which agriculture operates. These factors encompass a wide range of elements, including population growth, urbanization, income distribution, social inequalities, and cultural preferences. Understanding the interplay between agriculture and socioeconomic factors is essential for analysing the impacts of agricultural dynamics on food security, rural livelihoods, and overall economic development.

i. *Population Growth*: Population growth is a significant driver of agricultural dynamics. Increasing population puts pressure on agricultural systems to produce more food to meet the growing demand. Higher population densities can lead to increased competition for land, water, and other resources, which may necessitate changes in agricultural practices, intensification of production, or expansion of agricultural land.

ii. *Urbanization*: Urbanization, the process of population concentration in urban areas, has profound effects on agricultural dynamics. As people migrate from rural to urban areas, agricultural labour forces may decline, leading to changes in the structure of agricultural production. Urbanization also affects consumer preferences and demand for agricultural products, influencing the types of crops grown, the scale of production, and marketing strategies.

iii. *Income Distribution and Poverty*: Income distribution and poverty levels influence agricultural dynamics and rural livelihoods. Unequal distribution of income can result in disparities in access to resources, including land, credit, and inputs, which affect smallholder farmers' productivity and profitability. Poverty can limit farmers' capacity to adopt modern technologies, invest in their farms, and access markets. Addressing income inequalities and poverty is crucial for fostering inclusive agricultural growth and rural development.

iv. *Social Inequalities and Gender Dynamics*: Social inequalities, including gender disparities, have implications for agricultural dynamics. Women play a vital role in agriculture, yet they often face unequal access to resources, decision-making power, and control over land and assets. Addressing gender inequalities in agriculture can enhance productivity, improve food security, and contribute to more equitable agricultural systems.

v. *Cultural Preferences and Dietary Patterns*: Cultural preferences and dietary patterns influence agricultural dynamics by shaping the demand for specific agricultural products. Preferences for certain crops, livestock, or traditional foods impact farmers' choices and the types of agricultural systems that are dominant in a given region. Understanding cultural preferences and dietary habits is important for designing agricultural strategies that meet nutritional needs and promote sustainable food systems.

The impacts of socioeconomic factors on agricultural dynamics are multi-faceted:

i. *Food Security*: Socioeconomic factors influence food security by affecting the availability, accessibility, and affordability of food. Population growth, income distribution, and urbanization impact food demand and production systems. Ensuring food security requires strategies that consider these factors, such as promoting sustainable agriculture, improving rural infrastructure, and implementing social safety nets.

ii. *Rural Livelihoods*: Agricultural dynamics have direct implications for rural livelihoods, particularly in agrarian economies. Changes in agricultural practices, markets, and policies can affect farmers' incomes, employment opportunities, and quality of life. Supporting rural livelihoods requires addressing socioeconomic challenges, such as income inequality, access to resources, and diversification of rural economies.

iii. *Economic Development*: Agriculture is a significant sector in many economies, particularly in developing countries. Agricultural dynamics influence overall economic development through their contributions to employment, income generation, trade, and economic growth. Policies that promote agricultural productivity, value addition, and market access contribute to broader economic development goals.

iv. *Social Cohesion*: Socioeconomic factors influence social cohesion in rural communities. Inequitable access to resources, land conflicts, and social inequalities can undermine social stability and cohesion. Policies that promote social inclusion, equitable resource distribution, and participatory decision-making can contribute to more cohesive and resilient rural communities.

2.4.7 Resilience and Adaptation

Resilience and adaptation are critical concepts in understanding agricultural dynamics. Agriculture operates in a dynamic and unpredictable environment, and the ability of farmers and agricultural systems to adapt and build resilience is essential for sustainable production and livelihoods. This involves the capacity to respond and recover from shocks, such as climate variability, market fluctuations, policy changes, and other unforeseen challenges. Understanding the principles of resilience and adaptation helps in analysing the factors that drive agricultural dynamics.

i. *Climate Variability and Change*: Agriculture is highly vulnerable to climate variability and climate change. Changes in temperature, rainfall patterns, extreme weather events, and shifts in growing seasons can significantly impact agricultural production and productivity. Farmers need to adapt their agricultural practices, such as adjusting planting dates, selecting resilient crop varieties, implementing water management techniques, and adopting climate-smart agricultural practices, to cope with the changing climatic conditions.

ii. *Market Fluctuations*: Agricultural markets are subject to fluctuations due to various factors, including global supply and demand, trade policies, and market shocks. Farmers need to be adaptable in responding to price volatility, changing consumer preferences, and market demands. Diversifying crops, adopting value-added production, establishing alternative market channels, and engaging in cooperative marketing strategies are some ways farmers can adapt to market fluctuations.

iii. *Policy Changes*: Agricultural dynamics are influenced by changes in policies and regulations at various levels, including national, regional, and international. Farmers and agricultural systems must adapt to new policy frameworks, such as changes in subsidies, trade regulations, environmental standards, or land tenure systems. Flexibility in adjusting production methods, complying with new regulations, and seeking opportunities within new policy frameworks is crucial for the sustainability of agricultural systems.

iv. *Technological Innovation*: Technological innovation plays a vital role in enhancing the resilience and adaptation of agriculture. The development and adoption of new technologies, such as precision farming tools, biotechnology, remote sensing, and data analytics, enable farmers to better monitor and manage their farms, optimize resource use, and respond to changing conditions. Access to information and knowledge about innovative practices is essential for farmers to adapt and build resilience.

v. *Knowledge and Learning*: Resilience and adaptation in agriculture require continuous learning and knowledge exchange. Farmers, researchers, extension services, and other stakeholders need to collaborate to generate and share knowledge about best practices, innovative solutions, and lessons learned. Building adaptive capacity through training programs, farmer field schools, and knowledge-sharing platforms facilitates the adoption of new practices and enhances the ability of agricultural systems to cope with challenges.

vi. *Social Networks and Institutions*: Social networks and institutions play a crucial role in building resilience and facilitating adaptation in agriculture. Strong social networks enable farmers to access information, share resources, and collaborate during times of crises. Effective institutions, such as agricultural extension services, cooperatives, and farmer organizations, provide support, resources, and collective action platforms for farmers to adapt and respond to changing conditions.

vii. *Ecosystem Services and Biodiversity*: Ecosystem services and biodiversity play a vital role in enhancing the resilience and adaptive capacity of agricultural systems. Protecting and restoring natural habitats, promoting agro ecological practices, and maintaining biodiversity contribute to the resilience of ecosystems and agricultural landscapes. Ecosystem services, such as pollination, natural pest control, soil fertility, and water regulation, provide essential support to agricultural productivity and adaptation.

viii. *Risk Management and Insurance*: Risk management and insurance mechanisms help farmers cope with uncertainties and shocks. Crop insurance, weather-indexed insurance, and other risk-sharing mechanisms provide financial protection against crop failure, extreme weather events, and market risks. These mechanisms enable farmers to recover from losses and invest in resilience-building measures.

Building resilience and fostering adaptation in agriculture require a multi-faceted approach that combines technological innovation, policy support, knowledge exchange, social networks, and ecosystem management. It is important to empower farmers with the necessary resources, information, and skills to adapt to changing conditions, while also addressing broader systemic challenges such as climate change, market volatility, and policy constraints. By enhancing resilience and adaptation, agricultural systems can sustainably meet the evolving needs of food security, livelihoods, and environmental stewardship.

2.4.8 Knowledge and Innovation Systems

Knowledge and innovation systems play a crucial role in shaping agricultural dynamics. These systems encompass a wide range of actors, institutions, and mechanisms that generate, share, and apply knowledge and innovations within the agricultural sector. They contribute to advancements in agricultural practices, technologies, and approaches, leading to improved productivity, sustainability, and resilience. Understanding the dynamics of knowledge and innovation systems helps in analysing the factors that drive agricultural change.

i. *Research and Development (R&D)*: Agricultural research and development activities are at the core of knowledge and innovation systems. R&D

institutions, such as agricultural research centres, universities, and private companies, undertake scientific studies and experimentation to generate new knowledge, technologies, and best practices. This includes research on crop improvement, soil management, pest control, livestock production, and other areas relevant to agriculture. R&D efforts contribute to the development of innovative solutions and drive changes in agricultural practices.

ii. *Extension Services*: Extension services bridge the gap between research institutions and farmers by disseminating knowledge and information. Extension agents provide technical advice, training, and capacity-building support to farmers, enabling them to adopt new technologies, improve practices, and enhance their productivity. Extension services also play a role in facilitating knowledge exchange, farmer-to-farmer learning, and the adoption of innovative approaches.

iii. *Knowledge Exchange Platforms*: Knowledge exchange platforms facilitate the sharing and dissemination of agricultural knowledge and innovations. These platforms include conferences, workshops, seminars, farmer field days, and online networks. They provide opportunities for researchers, farmers, policymakers, and other stakeholders to interact, exchange ideas, showcase innovations, and learn from each other's experiences. Knowledge exchange platforms enhance collaboration, foster innovation, and contribute to the diffusion of new practices and technologies.

iv. *Farmer Networks and Participatory Approaches*: Farmer networks and participatory approaches promote farmer engagement and co-learning in the innovation process. Participatory research and extension methodologies involve farmers in the design, implementation, and evaluation of agricultural interventions. By actively involving farmers, their knowledge, needs, and perspectives are integrated into the development and adaptation of innovations. Farmer networks also serve as platforms for peer-to-peer learning, knowledge sharing, and collective action.

v. *Access to Information and Technology*: Access to information and technology is crucial for agricultural innovation. Farmers need timely and relevant information on market trends, weather forecasts, pest and disease management, and best agricultural practices. Information and communication technologies (ICTs), such as mobile applications, online platforms, and rural radio programs, have expanded access to agricultural information and enabled farmers to connect with knowledge and innovation systems more easily.

vi. *Institutional Support and Policy Environment*: Institutional support and a conducive policy environment are essential for effective knowledge and innovation systems in agriculture. Supportive policies and investments in

R&D, extension services, and knowledge exchange platforms facilitate the generation, dissemination, and adoption of agricultural innovations. Collaborative partnerships between research institutions, government agencies, non-governmental organizations, and private sector entities enhance the functioning of these systems.

vii. *Continuous Learning and Feedback Mechanisms*: Knowledge and innovation systems in agriculture require continuous learning and feedback mechanisms. Monitoring and evaluation of innovations and their impact, farmer feedback loops, and adaptive management approaches allow for ongoing improvements and refinements in agricultural practices. Learning from successes and failures, as well as incorporating feedback from farmers and other stakeholders, contributes to the evolution and effectiveness of knowledge and innovation systems.

viii. *Intellectual Property Rights*: Intellectual property rights (IPRs) play a role in incentivizing innovation and protecting the rights of innovators. Patents, plant variety protection, and other forms of IPRs provide legal protection for agricultural innovations, encouraging investment in research and development. Balancing the need for IPRs with ensuring access to knowledge and technologies for smallholder farmers and marginalized communities is crucial for equitable agricultural development.

Effective knowledge and innovation systems in agriculture facilitate the diffusion of new technologies, promote sustainable practices, and empower farmers to adapt to changing conditions. By promoting collaboration, learning, and the application of knowledge, these systems drive agricultural dynamics and contribute to the transformation of the agricultural sector.

2.4.9 Interdisciplinary Approaches

Understanding agricultural dynamics requires interdisciplinary approaches that transcend the boundaries of single disciplines and integrate knowledge from various fields. Agriculture is a complex system that encompasses ecological, social, economic, and policy dimensions. By adopting interdisciplinary approaches, researchers and stakeholders can gain a more comprehensive understanding of the complexities involved and develop holistic solutions to address agricultural challenges. Here are some key aspects of interdisciplinary approaches in studying agricultural dynamics:

i. *Integration of Multiple Disciplines*: Interdisciplinary approaches in agriculture involve the integration of knowledge and methodologies from multiple disciplines. Disciplines such as agronomy, economics, sociology, ecology, anthropology, geography, and policy studies each contribute unique insights and perspectives to the study of agricultural dynamics. Integrating these diverse disciplines allows for a more holistic understanding of the interactions and interdependencies within the agricultural system.

ii. *Systems Thinking*: Interdisciplinary approaches encourage systems thinking, which involves understanding agriculture as a complex and interconnected system. Rather than focusing on isolated components, systems thinking considers the relationships, feedback loops, and emergent properties of the agricultural system. This approach helps to identify the underlying drivers, dynamics, and potential unintended consequences of agricultural activities and interventions.

iii. *Holistic Problem-Solving*: Agricultural dynamics often involve multifaceted problems that require holistic solutions. By drawing on insights from different disciplines, interdisciplinary approaches facilitate the development of comprehensive and integrated strategies to address these complex challenges. For example, addressing soil degradation may require combining agronomic practices with socioeconomic interventions to promote sustainable land management and improve farmers' livelihoods.

iv. *Data Integration and Analysis*: Interdisciplinary approaches in agriculture involve the integration and analysis of data from various sources and disciplines. This may include agricultural statistics, climate data, socioeconomic indicators, satellite imagery, and qualitative data from surveys or interviews. By combining different types of data, researchers can gain a more nuanced understanding of agricultural dynamics and identify patterns, trends, and relationships that may not be apparent from a single disciplinary perspective.

v. *Stakeholder Engagement*: Interdisciplinary approaches emphasize the involvement of diverse stakeholders, including farmers, policymakers, scientists, and local communities. By engaging stakeholders from different sectors and disciplines, interdisciplinary research can better capture the complexities and nuances of agricultural dynamics on the ground. Stakeholder engagement also facilitates knowledge exchange, co-learning, and the co-creation of solutions that are contextually relevant and socially acceptable.

vi. *Policy and Practice Relevance*: Interdisciplinary approaches aim to generate knowledge that is directly relevant to policy and practice. By integrating perspectives from different disciplines, interdisciplinary research can provide policymakers and practitioners with comprehensive and actionable insights. This can inform evidence-based decision-making, the development of effective policies, and the implementation of sustainable agricultural practices.

vii. *Transdisciplinary Collaboration*: In addition to interdisciplinary approaches, transdisciplinary collaboration involves close collaboration between researchers and stakeholders across disciplines, including policymakers, farmers, non-governmental organizations, and the private sector. Transdisciplinary research integrates the knowledge and expertise of different stakeholders throughout the research process, fostering co-creation of knowledge, shared learning, and joint problem-solving.

viii. *Complex Problem Identification and Analysis*: Interdisciplinary approaches are particularly well-suited for addressing complex agricultural problems that transcend disciplinary boundaries. These problems often involve intricate interactions between ecological, social, economic, and policy factors. Interdisciplinary research helps identify the underlying causes, feedback loops, and trade-offs associated with such problems, enabling a more nuanced understanding and effective problem-solving.

By adopting interdisciplinary approaches, researchers and stakeholders can unlock new insights, develop innovative solutions, and promote sustainable agricultural development. Interdisciplinary collaboration enhances the capacity to address the multifaceted challenges of agricultural dynamics and fosters a more comprehensive understanding of the interactions between agriculture and its broader societal and environmental contexts.

This theoretical orientation provides a comprehensive framework for analysing and understanding the multifaceted factors that shape agricultural dynamics. By considering the interplay of technological advancements, environmental factors, market forces, policies, socioeconomic factors, resilience, knowledge systems, and interdisciplinary approaches, researchers and policymakers can gain valuable insights into agricultural dynamics and make informed decisions to foster sustainable agricultural development.

2.5 Mental Modelling and Fuzzy Logic Cognitive Mapping

In recent years, the application of cognitive science theories and methodologies in various fields has gained significant attention. One such field is agriculture and its allied sectors, which play a crucial role in global food production and sustainability. Mental modelling and fuzzy logic cognitive mapping are two theoretical frameworks that can be utilized to enhance decision-making processes and optimize resource allocation in these sectors. This theoretical orientation explores the potential applications of mental modelling and fuzzy logic cognitive mapping in agriculture and its allied sectors and highlights their benefits and implications.

2.5.1 Mental Modelling

Mental modelling refers to the cognitive process through which individuals construct internal representations of the external world and its dynamics. It involves the creation of mental models that capture relationships, structures, and patterns within a particular domain. In the context of agriculture and allied sectors, mental modelling can assist in understanding complex systems, predicting outcomes, and making informed decisions.

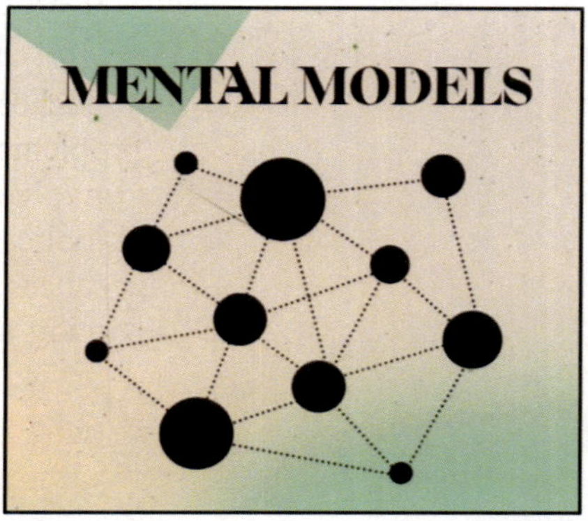

Fig. 20: Mental Models help deeper understanding on interaction and interdependencies.

2.5.1.1 Conceptualizing Mental Models in Agriculture

In agriculture, mental models can be developed to represent various components such as crop growth, pest management, irrigation systems, soil quality, weather patterns, and market dynamics. These mental models can help farmers, policymakers, and researchers to gain a deeper understanding of the interdependencies and interactions within the agricultural ecosystem.

Benefits of Mental Modelling in Agriculture

- *Improved decision-making*: Mental models enable stakeholders to simulate different scenarios and evaluate the potential outcomes of alternative agricultural practices, resource allocations, or policy interventions.
- *Enhanced problem-solving*: By representing the complex relationships and feedback loops, mental models can facilitate the identification of bottlenecks, vulnerabilities, and potential solutions in agricultural systems.
- *Knowledge transfer*: Mental models can serve as a communication tool, allowing experts to share their knowledge and insights with farmers, policymakers, and other stakeholders, leading to more informed discussions and collaborative decision-making.

2.5.1.2 Fuzzy Logic Cognitive Mapping

Fuzzy logic cognitive mapping (FCM) is an extension of mental modelling that incorporates fuzzy logic principles to capture the inherent uncertainties and vagueness in human cognition. FCM provides a flexible and intuitive framework for representing and analysing complex systems, allowing for the integration of qualitative and quantitative information.

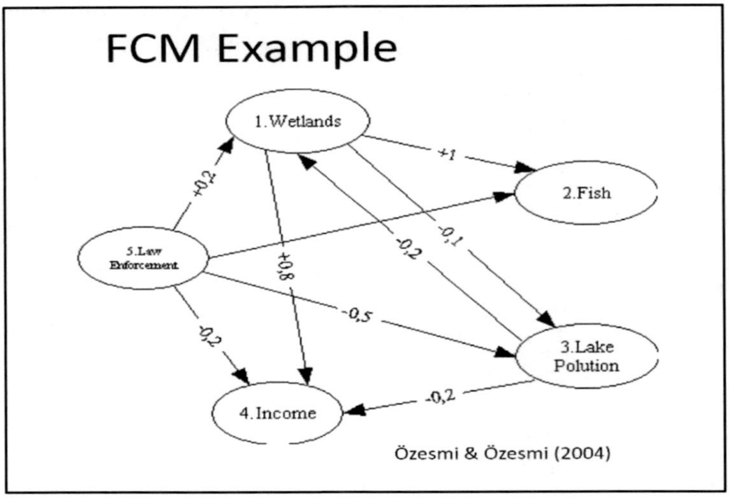

Fig. 21: FCM: Fuzzy Cognitive Mapping

2.5.2 Fuzzy Logic in Agriculture and Allied Sectors

The application of fuzzy logic in agriculture enables the consideration of imprecise and uncertain information related to crop growth, pest infestation, soil fertility, and other variables. FCM can incorporate expert knowledge, sensor data, and historical records to create a comprehensive model that captures the fuzzy relationships and dependencies among different factors in the agricultural domain.

Benefits of Fuzzy Logic Cognitive Mapping in Agriculture

- *Uncertainty handling*: FCM can handle imprecise and uncertain data, providing a more realistic representation of the inherent uncertainties in agricultural systems.
- *Adaptive decision-making*: Fuzzy logic allows for adaptive decision-making by accommodating changing environmental conditions and evolving knowledge.
- *Early warning systems*: FCM can be employed to develop early warning systems for pest outbreaks, crop diseases, and extreme weather events, enabling timely interventions and mitigation strategies.
- *Risk assessment*: By considering fuzzy relationships between variables, FCM can aid in assessing and managing risks associated with agricultural practices, resource allocation, and climate change impacts.

2.5.3 Integration of Mental Modelling and Fuzzy Logic Cognitive Mapping

The integration of mental modelling and fuzzy logic cognitive mapping can provide a comprehensive approach to understanding and managing agricultural systems. By combining the strengths of both frameworks, stakeholders can benefit from a more nuanced and accurate representation of the complexities involved.

2.5.3.1 Capturing Cognitive and Fuzzy Aspects

Mental modelling captures the cognitive processes and individual perspectives, allowing for a deeper understanding of how individuals perceive and reason about agricultural systems. Fuzzy logic cognitive mapping, on the other hand, incorporates the fuzzy nature of data and uncertainties, providing a more robust and flexible representation of the system dynamics. The integration of these two approaches enables a holistic view that considers both cognitive aspects and uncertainties in agricultural decision-making.

2.5.3.2 Leveraging Domain Experts and Stakeholder Involvement

The successful application of mental modelling and fuzzy logic cognitive mapping in agriculture and its allied sectors requires the active involvement of domain experts and stakeholders. Engaging farmers, researchers, policymakers, and other stakeholders ensures that the mental models and fuzzy logic maps accurately capture their knowledge, experiences, and concerns. Collaborative efforts can lead to more effective decision-making processes and the development of context-specific solutions.

2.5.3.3 Data-Driven Approach

The integration of mental modelling and fuzzy logic cognitive mapping can benefit from data-driven approaches. By incorporating real-time sensor data, historical records, and other relevant datasets, the models can be continuously updated and calibrated. Data-driven approaches provide a more empirical basis for decision-making, enhancing the accuracy and reliability of the models.

2.5.4 Potential Applications

Sustainable Agriculture: Mental modelling and fuzzy logic cognitive mapping can contribute to the development and implementation of sustainable agricultural practices. By considering the complex interplay between environmental factors, crop management, and socio-economic aspects, these frameworks can assist in optimizing resource allocation, reducing environmental impacts, and enhancing the overall sustainability of agricultural systems.

i. *Climate Change Adaptation and Mitigation*: Agriculture is highly vulnerable to climate change impacts, such as extreme weather events and shifting temperature patterns. Mental modelling and fuzzy logic cognitive mapping can aid in assessing the risks associated with climate change and designing adaptive strategies. These frameworks can support the identification of climate-resilient agricultural practices, crop selection, and land-use planning.

ii. *Market Analysis and Decision Support*: Mental modelling and fuzzy logic cognitive mapping can be utilized to analyse market dynamics, consumer preferences, and supply-demand relationships in the agricultural sector. By incorporating market data and economic factors, these frameworks can provide decision support for farmers and policymakers in optimizing production, pricing, and distribution strategies.

iii. *Agricultural Policy Formulation*: The integration of mental modelling and fuzzy logic cognitive mapping can assist policymakers in formulating evidence-based agricultural policies. These frameworks enable the evaluation of policy options, prediction of their potential impacts, and assessment of trade-offs between different objectives, such as food security, environmental sustainability, and socio-economic development.

2.5.5 Ethical and Social Implications

The application of mental modelling and fuzzy logic cognitive mapping in agriculture and its allied sectors raises important ethical and social implications that need to be carefully considered

- *Privacy and Data Security*: The use of these frameworks often requires the collection and analysis of sensitive data, such as farm management practices, financial information, and market data. It is crucial to ensure data privacy and security, protecting the confidentiality of stakeholders' information and preventing unauthorized access or misuse.
- *Equity and Access*: Equal access and distribution of the benefits derived from mental modelling and fuzzy logic cognitive mapping should be prioritized. Efforts should be made to bridge the digital divide, ensuring that farmers, especially those in marginalized or resource-constrained communities, have access to the necessary technology, training, and support to adopt and benefit from these frameworks.
- *Power Dynamics and Decision-Making*: The implementation of mental modelling and fuzzy logic cognitive mapping should consider power dynamics within the agricultural sector. It is essential to avoid reinforcing existing inequities and ensure that decision-making processes involving these frameworks are inclusive, transparent, and participatory, involving all relevant stakeholders.
- *Environmental Impact*: While mental modelling and fuzzy logic cognitive mapping aim to enhance sustainability, it is important to consider the potential environmental impacts associated with their implementation. The use of technology and data-driven approaches should be balanced with environmental stewardship, minimizing resource consumption, and considering the ecological consequences of agricultural practices.

2.5.6 Challenges and Considerations

While mental modelling and fuzzy logic cognitive mapping hold great potential in agriculture and its allied sectors, there are several challenges and considerations that need to be addressed for their successful implementation

- *Data Availability and Quality*: One of the key challenges is the availability and quality of data required for constructing and validating the models. Accurate and reliable data on various agricultural parameters, such as weather, soil quality, crop performance, and market trends, are essential for developing robust mental models and fuzzy logic maps. Efforts should be made to improve data collection systems and ensure data integrity.
- *Model Validation and Calibration*: Validating and calibrating the mental models and fuzzy logic maps can be complex due to the inherent uncertainties and subjective nature of the frameworks. It is crucial to establish validation procedures and involve domain experts and stakeholders in the validation process to ensure the accuracy and reliability of the models.
- *User Acceptance and Adoption*: For the successful implementation of mental modelling and fuzzy logic cognitive mapping in agriculture, user acceptance and adoption are critical. Farmers, policymakers, and other stakeholders need to understand the benefits and value of these frameworks and be willing to incorporate them into their decision-making processes. Efforts should be made to provide training, raise awareness, and address any concerns or resistance.
- *Complexity and Interpretability*: The complexity of mental models and fuzzy logic maps can pose challenges in terms of interpretation and understanding. Efforts should be made to simplify the representation and visualization of the models to facilitate effective communication and decision-making. Providing user-friendly interfaces and visualizations can enhance the usability and accessibility of these frameworks.

2.5.7 Future Directions

The application of mental modelling and fuzzy logic cognitive mapping in agriculture and its allied sectors is still an emerging field with immense potential for growth and development. Future research and advancements in this area can focus on the following directions

- *Integration of Machine Learning and Artificial Intelligence*: Leveraging machine learning and artificial intelligence techniques can enhance the capabilities of mental modelling and fuzzy logic cognitive mapping. The integration of these technologies can improve data analysis, pattern recognition, and prediction accuracy, enabling more sophisticated and dynamic models.

- *Participatory Approaches*: Further emphasis should be placed on participatory approaches, involving stakeholders at all stages of model development and implementation. Engaging farmers, policymakers, researchers, and other stakeholders in collaborative decision-making processes can lead to more context-specific and socially inclusive outcomes.
- *Scaling and Adaptation to Diverse Agricultural Systems*: Efforts should be made to scale up the application of mental modelling and fuzzy logic cognitive mapping to diverse agricultural systems, considering variations in climate, soil conditions, socio-economic contexts, and cultural factors. Adapting these frameworks to different contexts can unlock their potential in addressing global agricultural challenges.
- *Policy and Institutional Support*: To facilitate the adoption of mental modelling and fuzzy logic cognitive mapping in agriculture, policy and institutional support are crucial. Governments, research institutions, and international organizations can play a significant role in promoting the use of these frameworks by providing funding, technical assistance, and creating platforms for knowledge sharing and collaboration.

Mental modelling and fuzzy logic cognitive mapping offer promising avenues for improving decision-making processes and understanding complex agricultural systems. The integration of these frameworks in agriculture and its allied sectors can enhance sustainability, resource optimization, and resilience in the face of changing environmental conditions. However, it is essential to acknowledge the limitations and challenges associated with data availability, model validation, and user acceptance. Further research and practical implementation are necessary to unlock the full potential of mental modelling and fuzzy logic cognitive mapping in agriculture and its allied sectors.

3

Review and Citations

The Review of literature is a systematic, explicit and reproducible method of identifying, evaluating and synthesizing the existing body of a completed and a recorded work produced by researchers, scholars and practitioners. The Review of literature is also a resource paper which includes the current knowledge including substantive findings as well as the theoretical and the methodological contributions to a particular topic. These are available mostly in academic and scholastic journals.

Review of literature provide the basics for research in nearly every academic field. This chapter is an extensive survey of all past studies relevant to the field of investigation which gives knowledge of what others have found out in the related field of study. In any scientific investigation or research, a comprehensive presentation of the review of literature is a statutory requirement.

The aim of review of literature is to highlight what has been done so far in the field of interest and how our findings are related to earlier research.

Its main functions are to develop a better understanding of the problem to be investigated, delineate a new area of study and avoid unnecessary duplication, decide on the tools and techniques to be adopted including developing some new ones and also to relate the present study with the previous ones by finding out the areas of agreement and disagreement.

The review of literature may

(a) Indicate the following by providing referential inputs – Approaches, Methods, Variables used, Statistical procedure, Expected results.

(b) Provide theoretical background to the study.

(c) Help to justify how the findings are related to the body of knowledge in the field of research.

(d) Establish a link between what we have proposed to study and what has already been found. It helps us to refine the research methodology.

The literature review in this study focuses on the key concepts relevant to the field of investigation. Given the limited availability of direct references on the chosen topic, this chapter includes specific references as well as indirect sources to ensure meaningful utilization of information.

Accordingly, a brief review of the available literature has been incorporated in the light of the objectives of study under the following heads:

i. Climate Change and Global Warming
ii. Uncertainty and Risk
iii. Change dynamics in agriculture
iv. Estimation of perception and evaluation of interactions between variables
v. Mental Modelling and Fuzzy-logic Cognitive Mapping

3.1. Climate Change and Global Warming

As on Wikipedia (2022), Climate change is adversely affecting both terrestrial and marine ecosystems and is anticipated to continue its impact on numerous ecosystems in the future. The consequences of climate change, such as the escalation of global temperatures, extreme weather occurrences, and rising sea levels, are resulting in the decline and extinction of species, alterations in ecosystem behaviour, and an increase in the prevalence of invasive species. As per the IPCC Sixth Assessment Report, extreme weather events are projected to become more frequent and intense due to the escalating global warming. The implications of climate change for biodiversity encompass the potential occurrence of a new mass extinction event, putting over 50% of animal and plant species at risk of eradication. Numerous species at risk are found within the Arctic and Antarctic regions, including polar bears and emperor penguins. Additionally, the rising temperatures are instigating modifications in the behaviour of birds and butterflies, as well as prompting shifts in their geographical ranges towards the poles or higher elevations.

Ortiz et al. (2021) discusses the intricate connections among biodiversity, agriculture, climate change, and global trade are examined in this article. As the world's population continues to grow, it becomes crucial to discover sustainable methods of feeding everyone while safeguarding the environment and biodiversity. The most significant threat to biodiversity arises from changes in agricultural land use, which are further intensified by the impacts of climate change. The international trade of agricultural goods also plays a role in the loss of biodiversity, particularly in developing nations. To address these trade-induced consequences, the article emphasizes the necessity of comprehending the interactions between international trade, production, and biodiversity, thereby enabling the design of evidence-based policies and programs. The article outlines eight key priorities that will enhance our understanding of the complexities within the environment-agriculture-trade system.

Malhi et al. (2020) discusses the impact of climate change on ecosystems and emphasizes the importance of comprehending the ecological dynamics associated with these impacts. It also explores the potential of nature-based solutions for mitigating and adapting to climate change. The article introduces a thematic issue that focuses on the interaction between climate change and the biosphere, and it identifies key priorities for both academic research and practical implementation in order to preserve a diverse, resilient, and well-functioning biosphere in the challenging conditions of the twenty-first century. Additionally, the article provides a summary of the findings from a Royal Society-National Academy of Sciences Forum that took place in Washington DC in November 2018. This forum delved into the latest scientific knowledge regarding how climate change affects terrestrial, freshwater, and marine ecosystems. The forum specifically examined the effects of climate variability and extremes, the interactions between climate change and other human-induced stressors, the existence of thresholds and the potential for sudden and irreversible change, and the interactions between different trophic levels in ecosystems. Finally, the article concludes by highlighting the necessity for further research to better understand the threats, opportunities, and solutions concerning climate change and ecosystems.

Weiskopf et al. (2020) discussesthe consequences of climate change on the variety of living organisms, ecosystems, and the services they provide in the United States. They utilize information from the 4th National Climate Assessment to summarize the observed and predicted alterations in ecosystems and biodiversity. Additionally, they investigate the connections to significant ecosystem services while discussing the difficulties and opportunities linked to the management of natural resources. It is revealed that species are adapting to climate change by modifying their physical characteristics, behaviour, timing of natural events, and their geographical distribution. These modifications occur through both flexible and long-term evolutionary responses. Such alterations have an impact on the services provided by ecosystems and the well-being of human beings. Consequently, natural resource managers must adopt proactive and adaptable approaches to effectively cope with these transformations.

Moore et al. (2017) discusses the quantification of the difference between different approaches in assessing the impacts of climate change on crop yields. The paper uses a database of studies compiled for the IPCC 5th Assessment Report to analyse the impact of temperature, rainfall, and CO2 concentrations on the yield of maize, rice, soy, and wheat. The database includes over 1700 point estimates from 94 different studies, incorporating both process-based crop models and empirical papers. The authors focus on these four crops as they are extensively studied and account for a significant portion of global agricultural production. The database also includes information on on-farm adaptations, such as adjusting planting dates or cultivars. The authors complement the existing database by coding each study based on the approach used (process-based or empirical) and adding baseline growing-season temperatures. They treat the database as an "ensemble of opportunity," drawing from a set of possible models to capture the response of crops to changing climate conditions. Multiple regression analysis is used to aggregate the results from individual studies and estimate common response functions. The regression model includes variables for changes in temperature, CO_2 concentration, rainfall, and other factors that may influence crop yields.

Mishra and Sahu (2014) discusses the use of the Ricardian approach to measure the economic impact of climate change on the agriculture sector, specifically in the coastal zone of Odisha, India. The study estimates a functional relationship between farm-level net revenue and climate variables (temperature and rainfall) using regression analysis. The data used includes monthly temperature and rainfall data from 1979-2009, as well as annual district-level agricultural data and other control variables. The study finds that climate change is likely to have a negative impact on agriculture in the region. The study also provides the regression coefficients for the climate variables and control variables used in the analysis.

Grimm et al. (2013) discussesthe effects of climate change on ecosystems in the United States. They provide a summary of key findings, focusing on the impact of climate change on the movement of matter and energy within ecosystems, as well as the living and non-living components that play a significant role in these processes. The article identifies seven crucial impacts of climate change on the structure and functioning of ecosystems in the US. These include the reduction of sea ice, rapid warming, and increased organic inputs, which affect the productivity of marine and lake ecosystems. Additionally, there are shifts in species distribution that alter the composition of biomes across 5-20% of US land area. The article also explores the economic and human consequences resulting from changes in ecosystem productivity, food-chain dynamics, the spread of diseases, the transport of pollutants, and climate feedback mechanisms.

Bush et al. (2011) discusses a workshop held in Goa, India in 2009 to discuss the human health effects of climate change in India. The workshop was cosponsored by the University of Michigan's Centre for Global Health, the U.S. Centres for Disease Control and Prevention's National Centre for Environmental Health, and the Indian Council of Medical Research. The participants discussed the current state of the science, identified gaps in understanding, and outlined future research directions related to climate change and health in India. It highlights the diversity and vulnerability of India, as well as the challenges associated with conducting research and improving public health infrastructure in the face of climate change. It also discusses the prevalence of waterborne infectious diseases and the burden they place on the health and economic sectors in India. It emphasizes the need for region-specific research and prevention strategies, as well as the importance of building a data infrastructure and conducting monitoring in India. Heat stress and air pollution are also mentioned as significant health concerns in India.

EPA (2010) discusses the impacts of climate change on ecosystems, encompassing biodiversity, oceans, forests, habitats, invasive species, and natural phenomena such as migrations and life cycle events, are examined. Climate change can result in species extinction, habitat modifications, increased wildfires, and disturbances to natural patterns. The oceans absorb carbon dioxide from the atmosphere, leading to elevated levels of carbonic acid, which negatively affects marine life. Invasive species may thrive under these changing conditions, potentially displacing native species. The timing of natural events is intricately linked to climate factors, and alterations in weather patterns can disrupt seasonal behaviour and interactions among species. The passage emphasizes the crucial role of ecosystems, as they provide people with sustenance, resources, medications, and numerous other products, while also playing a vital role in nutrient recycling, water purification, and climate regulation.

A report authored by Nye (2010) on the condition of the Gulf of Maine has been published by the Gulf of Maine Council on the Marine Environment and the National Oceanic and Atmospheric Administration. The report provides an overview of how climate change impacts the ecosystems, habitats, and organisms within the Gulf of Maine. It emphasizes that human activities, particularly the escalating emissions of greenhouse gases, are driving changes in the Earth's climate, resulting in warming of both the atmosphere and the oceans. The report further examines the various factors contributing to climate change, including human-induced climate change, population and economic growth, and natural conditions. In the concluding section, the report addresses measures and responses, including government policies, action plans, and ongoing monitoring and research efforts.

Thomas et al. (2008) discusses the consequences of climate change on natural ecosystems and biodiversity, and its varying impact on different countries and regions. It is anticipated that the poorest nations, where people heavily depend on natural systems for their livelihoods, will encounter the most significant transformations. The article assesses the countries and regions that will be most severely affected in terms of changes to their natural ecosystems and biodiversity, and it also explores the global distribution of responsibility for these changes. Furthermore, the article highlights the significance of vegetation change as a crucial outcome of a shifting climate, as it directly influences the habitats of numerous terrestrial species. In summary, the alteration of vegetation will disproportionately affect countries that possess rich biodiversity.

Malcolm and Pitelka (2000) authored report, titled "Ecosystems and Global Climate Change" explores the potential impacts of climate change on terrestrial ecosystems and biodiversity in the United States. It emphasizes that global warming will bring about significant alterations in the distribution of ecosystems across the country. In the eastern part of the US, there will be a general shift northward in the types of vegetation, while the changes in the western US will be more intricate and localized. The report also recognizes that climate change will affect the functioning of ecosystems, leading to variations in plant productivity and rates of decomposition. It acknowledges that there are considerable uncertainties regarding the specific effects of climate change on ecosystems and biodiversity in the US. Moreover, the report suggests that the impact of climate change on ecosystems and species will likely be intensified in areas where human activities have already put pressure on these ecosystems.

3.2. Uncertainty and Risk

Kamali et al. (2022) examines the primary factors that lead to uncertainties in simulated yields of irrigated maize. It investigates these uncertainties under historical conditions and scenarios involving increased temperatures and changes in irrigation water availability. The research quantifies the relative impacts of three irrigation water allocation strategies, three sowing dates, and three maize cultivars on the uncertainty of the simulated yields. The water allocation strategies were determined based on historical data regarding farmers' allocation patterns in a drip-irrigation scheme in the Genil-Cabra region of Spain between 2014 and 2017. The findings indicate that irrigation strategies are the major driver of uncertainty in the simulated yields, accounting for 66% of the total uncertainty. However, when temperature increase scenarios are considered, the choice of crop model and cultivar becomes equally significant in contributing to the uncertainty in simulated yields. This is partly due to variations in the model structure related to temperature responses. The study emphasizes the importance of incorporating information on farmers' irrigation strategies to mitigate uncertainties in simulated yields at the field scale.

Patra et al. (2022) discusses the factors contributing to the uncertainty and vulnerability of Indian agriculture, including weather fluctuations, market volatility, limited access to technology, and an unreliable input delivery system. The paper explores various issues and potential reasons for this uncertainty, such as farmers' suicides, the impact of climate change on soil and water erosion and depletion, and the challenges associated with achieving sustainable development in Indian agriculture. However, the article lacks a comprehensive solution to address the uncertainty and vulnerability in Indian agriculture. Furthermore, it does not adequately address the role of government policies and regulations in reducing these uncertainties and vulnerabilities in the agricultural sector of India.

Mehta et al. (2021) written book titled "The Politics of Climate Change and Uncertainty in India" delves into the influence of uncertainty on climate and environmental policies within the context of India. Focusing specifically on India, a country highly vulnerable to climate impacts, the book presents case studies from various regions. It critically examines the diverse discourses, practices, and political dynamics surrounding uncertainty, highlighting significant variations in how different groups, categorized as "above," "middle," and "below," perceive and experience climate change and uncertainty. The central argument of the book is the need for inclusive and socially just approaches to address climate change not only in India but also on a global scale. By shedding light on the complex interactions between climate change, uncertainty, and politics, the book aims to contribute to pluralistic strategies for mitigating and adapting to climate change. It is part of the Pathways to Sustainability Series, which focuses on tackling core challenges related to connecting science and technology, environmental sustainability, poverty reduction, and social justice.

Shoaib et al. (2021) discusses the significance of comprehending the long-term factors that influence food security and how they interact with each other. Its purpose is to provide guidance to policymakers when formulating policies for future food security. The article also emphasizes the intricate and uncertain nature of multidimensional food security, highlighting the necessity for scenario analysis based on models to forecast future food security and facilitate decision-making. The research investigates the uncertainty associated with a hydrologic food security model that considers the impact of climate warming on food production in a study site with mild temperatures in China. Additionally, the study explores the effects of three different CO_2 emission scenarios on food production. The article presents an innovative and comprehensive approach to quantifying uncertainty in food security modelling and showcases the extent of uncertainty by utilizing crop yield data from China. Ultimately, the article concludes that there is an overall upward trend in rice, cereal, and wheat production under a warming climate. The variability of crop yields along with the temperature increase provides a fresh perspective on quantifying uncertainty in food security modelling.

Waldman et al. (2020) examines the use of heuristics in agricultural decision-making by smallholder farmers in developing countries, in response to climate uncertainty. The authors argue that traditional assumptions of rational, utility-maximizing approaches are inadequate for understanding decision-making in complex and uncertain environments. They propose that researchers studying climate uncertainty and agricultural decision-making should incorporate both rational and heuristic decision-making approaches. Several challenges are identified, including the lack of meteorological data and climate information in developing countries, variations in perceptions of climate change, and cognitive biases related to climate observations. These factors make it difficult for farmers to make optimal decisions. The article emphasizes the necessity for a more realistic decision-making framework that recognizes the prevalence of simple heuristics, cognitive biases, and the significance of climate compared to other risk factors. In summary, the article advocates for a nuanced understanding of decision-making processes in agriculture, considering the influence of heuristics, cognitive biases, and the limited availability of climate-related information in developing countries. By incorporating these factors into research and decision-support tools, a more accurate understanding of smallholder farmers' decision-making in the face of climate uncertainty can be achieved.

Mehta et al. (2019) addresses the significant role of uncertainty in shaping climate and environmental policies at different levels, ranging from global to local. Uncertainty refers to the inability to precisely determine outcomes, effects, or impacts of an event, where probabilities cannot be accurately calculated. As climate change progresses, there is an expected increase in both climatic variation and uncertainty, particularly as one moves from the global to local scale. Over the past two decades, there has been a shift in focus from reducing scientific uncertainty to comprehending and managing uncertainty as a whole. This special issue focuses on exploring the concept of uncertainty in relation to climate change from various perspectives. It examines different discourses and practices surrounding climate change uncertainty, considering both perspectives from those directly impacted by climate change ("below") and those involved in policymaking and governance ("above"). Furthermore, it investigates the interactions between these perspectives in diverse socio-ecological contexts, as well as the politics of knowledge that shape and influence responses to uncertainty. The authors analyse how differing understandings of uncertainty can either hinder or provide opportunities for adapting to climate change. They also explore approaches that can bridge the gaps between diverse perspectives of uncertainty, with the aim of advancing socially just pathways for adaptation.

Mehta, Srivastava, et al. (2019) examines the impact of climate-related uncertainty on various sectors in India. It introduces the heuristic framework of 'above', 'middle', and 'below' to analyse how different groups comprehend and experience climate change and uncertainty. Responses from actors categorized as 'above' predominantly focus on controlling uncertainty through top-down, techno-managerial solutions. Meanwhile, scientists rely on quantitative assessments and models based on probabilistic scenarios. However, these approaches often fail to capture the experiences and realities of local communities, particularly those in the global South who bear the brunt of climate change impacts. The article argues that embracing diverse knowledge and approaches is crucial for understanding and addressing climate-related uncertainties, ultimately promoting socially just adaptation. In conclusion, the article presents the conceptual framework and heuristic of 'above', 'middle', and 'below', along with the research methodology and a description of the research sites. It further explores the varied discourses and practices of uncertainty within these three settings before providing some final thoughts on the subject.

Asseng (2015)examines the uncertainties associated with predicting the effects of climate change on agriculture using crop simulation models. It highlights that variations in simulated climate change impacts among different agricultural impact models arise from differences in model structures and parameter values. The uncertainties in impact simulations are found to be more significant due to crop models rather than downscaled general circulation models. The article proposes that reducing impact uncertainties can be achieved by enhancing model routines and parameters through detailed field experiments. Additionally, the use of multi-model ensemble medians improves predictions compared to individual models. In conclusion, the article emphasizes the importance of reducing uncertainty in understanding how climate change will impact agricultural productivity to facilitate the development of effective adaptation strategies and policies.

Intergovernmental Panel on Climate Change, (2015) written Chapter 2 of the IPCC's Fifth Assessment Report examines the comprehensive assessment of risks and uncertainties associated with climate change response policies. The chapter is written by a team of experts representing various countries including the USA, India, Italy, France, Germany, Cuba, Austria, Switzerland, and Nigeria. It delves into several key areas such as evaluating uncertainty and risk through appropriate metrics, understanding how risk is perceived and how individuals respond to uncertainties, and exploring tools and decision aids for analysing and managing uncertainty and risk. The authors emphasize the significance of incorporating uncertainties into climate policy decisions and highlight the implications of intuitive decision-making processes. The chapter concludes by discussing the benefits and limitations of utilizing cost-benefit analysis to enhance decision-making procedures.

Noleppa, (2013) authored and The Deutsche Gesellschaftfür Internationale Zusammenarbeit (GIZ) published report focuses on economic approaches to evaluate climate change adaptation options amidst uncertainty. The report examines the necessity and lessons learned in employing economic approaches for assessing these options. It assesses various methodological approaches commonly used and considered most promising for economically evaluating climate change adaptation options. Evaluation criteria, including data requirements and availability, scaling, necessary skills, and compatibility with ci:grasp, are thoroughly reviewed. The report presents two prototypes of software or programming solutions based on Microsoft Excel, illustrating their practicality in assessing climate change adaptation options using real or hypothetical data. In conclusion, the report offers recommendations on integrating economic assessment activities into development projects and programs, as well as suggestions for further steps to be taken after this study.

Mastrandrea et al. (2010) written document titled "The Guidance Note for Lead Authors of the IPCC Fifth Assessment Report on Consistent Treatment of Uncertainties" offers guidance to authors on ensuring consistent treatment of uncertainties across all three Working Groups. The note establishes a unified approach and standardized language that can be widely utilized for developing expert judgments, as well as for assessing and effectively conveying the level of certainty in assessment findings. In the context of the AR5, two metrics will be employed to communicate the degree of certainty in key findings: confidence in the validity of a finding and quantified measures of uncertainty expressed probabilistically. The guidance notes specifically encourage author teams to provide information on the extreme ends of variable distributions, presenting quantitative estimates whenever feasible, and providing qualitative assessments and evaluations when appropriate. Additionally, the note offers guidance on addressing issues related to uncertainty, such as considering all plausible sources of uncertainty, evaluating uncertainty and risk factors, and communicating findings at the appropriate level of detail.

Quiggin, (2008) addresses the issue of uncertainty surrounding climate change and proposes how economists can incorporate it into their analysis. The uncertainty can be categorized into three stages: scientific uncertainty, impact uncertainty, and policy uncertainty. The scientific uncertainty is examined in the IPCC's Third Assessment Report (TAR), which identifies five types of uncertainty. Impact uncertainty refers to understanding how changes in the climate system result in significant consequences for humans. Policy uncertainty involves the uncertainty surrounding the future policies implemented to regulate emissions in the coming years. The article emphasizes the importance of considering all three stages of uncertainty when evaluating the economic aspects of climate change.

3.3. Change dynamics in agriculture

Benton et al. (2021) discusses the trends in biodiversity loss and the impacts of the food system on biodiversity. It highlights the importance of biodiversity for human well-being and the environment. Biodiversity is declining at an unprecedented rate, with the global rate of species extinction being tens to hundreds of times higher than the average rate over the past 10 million years. Around a quarter of species in most animal and plant groups are already under threat from extinction, and approximately 1 million more species face extinction within decades. The decline in biodiversity has negative consequences for the functioning of ecosystems and the services they provide, such as water supply, soil fertility enhancement, pollination, and natural pest control. Biodiversity also has positive impacts on quality of life, physical and psychological well-being, and mental health. Access to natural spaces and exposure to animal and plant species are associated with positive outcomes for well-being, even in urban settings. The literature emphasizes the need for transformative changes in the food system to support biodiversity and nature conservation.

Shi et al. (2020) discusses various studies conducted on the impact of climate change and spatial differences on agricultural production. Some studies analyse the relationship between climate change and agricultural production, while others examine regional differences in agricultural productivity. The studies use different methodologies such as production function models, climate trend analysis, and econometric models to assess the effects of climate variability and regional factors on agricultural production. The results highlight the negative impact of climate change, such as drought and extreme high temperatures, on grain yield. Additionally, the studies reveal regional heterogeneity in agricultural technology, energy efficiency, and productivity, with variations observed between different regions in China. The findings emphasize the need for effective decision-making and strategies to address climate-related challenges and improve agricultural production efficiency.

Arora (2019) explores the repercussions of climate change on agricultural production and proposes sustainable solutions. The rising global population's increasing food demand has led to intensive agricultural practices that, unfortunately, exacerbate the issue by releasing greenhouse gases and polluting natural resources. Climate change has accelerated land degradation, leading to heightened desertification and nutrient-depleted soils. To address these challenges, the article recommends adopting climate-smart agricultural systems, cultivating resilient crop varieties, implementing biotic and sustainable farming methods, and modifying crop cultivars, sowing times, and cultivation techniques to enhance crop resilience in extreme conditions. However, it is important to note that the article does not delve into the potential economic costs associated with implementing climate-smart agriculture systems and developing resilient crop varieties. Furthermore, it does not address the potential challenges of implementing these solutions on a global scale and the need for international cooperation.

Srinivasa Rao et al. (2019) discusses the impacts of climate change on various sectors in South Asia, particularly focusing on agriculture. It highlights the rising sea levels, which could lead to the submergence of the Maldives Island. Droughts are also a major concern, affecting regions in India and Pakistan, and leading to losses in economic activities and livelihoods. Extreme events such as cyclones and storms have severe consequences, with millions of people being affected. Heatwaves are becoming more frequent and intense, negatively impacting agriculture and food security. The melting of glaciers in the Himalayan region poses a threat to the availability of freshwater. The passage emphasizes that agriculture, being a significant occupation in the region, is highly dependent on climatic conditions, and disruptions in weather patterns can lead to a significant decrease in crop productivity. This can have severe implications for food security, particularly for resource-poor individuals and small farmers. The passage also includes a figure showing the cropped areas affected by natural extreme events in India over the years.

Mahato (2014) explores that climate change is expected to have various impacts on agriculture. Higher temperatures, changes in precipitation patterns, and increased atmospheric CO_2 concentrations are some of the key factors that will affect agricultural productivity. Higher CO_2 concentrations can lead to increased photosynthetic rates and reduced transpiration in plants. However, the response of plants to elevated CO_2 levels is influenced by various environmental and plant factors, making it difficult to predict the exact effects. Rice yields are expected to increase with higher CO_2 concentrations but decrease with higher temperatures. CO_2 enrichment has shown significant increases in rice biomass and yields at ambient temperatures, but these increases tend to be offset when temperature is increased along with rising CO_2. Spikelet sterility, caused by high temperatures, is a major factor contributing to yield losses in rice. High CO_2 levels and/or temperature are also likely to affect crop development rates. Other expected changes due to climate change include rising sea levels, changes in precipitation patterns, increased storminess, and increased variability in climatic variables. These changes can lead to land loss, coastal erosion, flooding, drought risk, soil erosion, and changes in the timing of farm operations. Climate change will also impact soil microbial processes, crop residue decomposition, and the prevalence of pests and diseases. Overall, climate change poses significant challenges to agricultural productivity and food security.

Nastis et al. (2012) explores the consequences of climate change on agricultural productivity, specifically examining its impact on Greece. The authors contend that agriculture stands as the economic sector most susceptible to climate change and emphasize the necessity for farmers to adapt to the shifting climate conditions. The paper analyses the economic repercussions of climate change on Greek agricultural productivity over the past three decades and deliberates on the implications for policymakers and agricultural research. The authors propose that policies should consider the multifaceted nature of modern agriculture and prioritize sustainable agricultural development. However, it is important to note that the article primarily focuses on Greece, and the influence of climate change on agricultural productivity may vary across different regions. Additionally, the article does not delve into the potential benefits of climate change on agriculture, such as extended growing seasons and increased levels of carbon dioxide that can enhance photosynthesis.

Gornall et al. (2010) discusses the impact of climate change on agricultural productivity. It highlights the negative effects of warming on crop yields, with estimates of annual combined losses of 40 million tonnes or US$5 billion since 1981. The largest warming is projected in high latitudes, but even small increases in temperature in low latitudes can have a significant impact due to already marginal agricultural conditions. Precipitation patterns also play a crucial role in agriculture, with over 80% of total agriculture being rain-fed. Projections of future precipitation changes influence the magnitude and direction of climate impacts on crop production. There is uncertainty in predicting regional precipitation changes, but there is increasing confidence in projections of increased precipitation in high latitudes and decreased precipitation in many parts of the tropics and sub-tropics. The article also discusses the influence of climate variability and extreme weather events on food security. Changes in year-to-year variability and extreme weather events pose greater risks to food production than long-term mean climate change. Historically, low precipitation events have led to significant falls in crop productivity. Even small changes in mean annual rainfall can impact productivity. The article concludes by emphasizing the need for ongoing adaptation to climate change and the importance of addressing climate variability and extreme weather events to ensure food security.

According to Aydinalp and Cresser (2008) the agriculture sector is a significant contributor to greenhouse gas emissions, particularly methane (CH_4) and nitrous oxide (N_2O). Methane emissions primarily come from paddy fields, animal husbandry, and burning agricultural wastes. Livestock and manure management contribute to 16% of total CH_4 emissions, with buffalo and cattle being the main sources. Nitrous oxide emissions mainly result from nitrogen fertilizer usage, legume cropping, and animal waste. Better application of nitrogen fertilizers and improved treatment of animal wastes can help reduce N_2O emissions. Energy use in agriculture has decreased since the 1970s, but further reductions can be achieved through practices like irrigation scheduling, minimum tillage, and improved fertilizer management. Forests and vegetation play a crucial role in carbon sequestration, with unmanaged forests holding significantly more carbon than agricultural lands. Climate change will have varying impacts on agriculture globally, with changes in temperature, rainfall patterns, and CO_2 levels affecting crop productivity, pests, and diseases. Regional impacts could be significant, especially in tropical regions. Vulnerability to climate change depends on physical, biological, and socioeconomic factors, with low-income populations in isolated agricultural systems being particularly vulnerable. Marginal agriculture and farmers at the edge of climatic regions are also at risk. The potential negative effects of climate change on agriculture include uncertainties in regional climate predictions and changes in crop yields. However, some regions may benefit from warming and increased CO_2 levels.

3.4. Estimation of Perception and Evaluation of Interactions Between Variables

Bairagya (2022) discusses the relationship between productivity, income, livelihood, and ecological resilience, noting that they often contradict each other. However, it is possible to reconcile these factors through synchronization. The study was conducted in Mitrapur, Haringhata block, Nadia district of West Bengal, India, with 75 respondents selected through snowball sampling. The research location has a modernization process involving farmers' participation and entrepreneurial institutionalization. The study measured the variables of productivity, livelihood, income, and ecological resilience using 16 exogenous variables. Data was collected between May and July 2022. The study identified certain variables such as age, education, family size, income, expenditure, homestead land, economic land, electricity consumption, fuel consumption, market interaction land, group interaction, distant matrix, innovation proneness, market orientation, biodiversity index, and soil conservation as strong determinants in reconciling productivity, livelihood, and ecological resilience. Factors like family size, homestead land, distant matrix, innovation proneness, and diversity index were found to have strategic implications for policy interventions to improve income, productivity, and livelihood.

Chandra (2022) discusses the importance of happiness for Indian farmers and how their well-being is often overlooked. It highlights the uncertainty and stress that farmers face due to unpredictable markets, income, and wages. The study mentions that India ranks low on the happiness index globally. The study aims to identify the factors that contribute to farmers' happiness by analysing data from 100 farmers and examining various variables. The variables found to have the most significant impact on happiness include cropping intensity, market variability, health status, income uncertainty, and weather vulnerability. The study concludes by emphasizing the need to understand the causes of farmers' unhappiness and implement strategies to improve their well-being.

Kundu (2022) discusses the importance of agriculture in India's economy and how many rural households depend on it. It highlights the need for innovation and improvement in agriculture, and how Agricultural Extension can help in developing and managing strategies. Communication is identified as a crucial tool in enterprise management, specifically in entrepreneurial communication. The objective of the study mentioned in literature is to estimate and analyse the process, factors, and impact of entrepreneurial communication in agricultural enterprises. The study selected 19 independent variables and 6 dependent variables, which were measured using scales developed by previous researchers. The village of Radhikapur in the Kaliyaganj block of West Bengal was chosen for the study, with 75 respondents participating. Various statistical tools were used for analysis, including mean, median, mode, standard deviation, correlation coefficient, regression analysis, path analysis, and factor analysis. The study found significant correlations between certain variables and the different aspects of entrepreneurial communication. The variables that had the most impact on the dependent variables were average family education, year of enterprise, annual income, family size, and family expenditure. The study concludes that most variables were positively and significantly correlated with entrepreneurial communication. It also mentions that the findings have policy implications, particularly in the context of West Bengal's social ecology.

Patra (2022) discusses the inherent uncertainty in agricultural production systems, particularly in the Indian context. The system is vulnerable to various uncertainties such as unpredictable weather, market volatility, limited access to technology, and inefficient input delivery. The study focuses on understanding and identifying the factors contributing to uncertainty by surveying 100 small and marginal farmers. The study finds that weather, income, wage, and access to the market are the most significant contributors to perceived uncertainty. Additionally, the study utilizes Artificial Neural Network (ANN) to analyse the relationship between variables. The findings have important policy implications for crop insurance, wage determination, minimum support price, and risk management in agriculture.

3.5. Mental Modelling and Fuzzy-logic Cognitive Mapping

Van Hulst et al. (2020) discussesthe disparities in the understanding of agro-ecology between scientists and farmers in Scotland. The authors employed a collaborative approach to construct mental models, aiming to identify areas of agreement and disagreement in their understanding. The findings reveal that scientists predominantly perceive agro-ecology in Scotland as a scientific field that applies ecological analysis to agricultural systems. On the other hand, farmers possess a broader perspective that encompasses the entire food system, encompassing aspects such as consumer health, markets, and sustainable energy. The article concludes that while there are significant differences between farmers and scientists, there are also areas of shared understanding. For instance, both groups acknowledge the potential of novel crops and new crop rotations, which can serve as a foundation for collaboration towards an agricultural approach that yields multiple benefits.

Friedrichsen et al. (2018) discusses emphasizes the significance of comprehending stakeholders' mental models regarding soil management for food security in peri-urban areas of India. The agricultural advancements of the Green Revolution in India contributed to achieving food sovereignty; however, food insecurity still persists. The focus on increasing crop production during that period neglected to consider the broader implications for food security. Scientists, extension agents, and farmers hold distinct perspectives on how soil health impacts food security. Mental models serve as a tool for understanding an individual's cognition, and effective communication is more likely to occur when there is alignment between their mental models. Participatory research and development play a vital role in fostering shared mental models. The article proposes the inclusion of a fifth dimension, cultural acceptability, in agricultural technology development and dissemination, highlighting its importance within the context of food security.

Levy et al. (2018) highlights the significance of systems thinking in promoting sustainability objectives, particularly in complex systems like sustainable agriculture. The authors propose a novel approach for evaluating systems thinking by utilizing social network tools to analyse mental models. They examine the cognitive maps of 148 influential figures in sustainable agriculture in California and assess the degree to which each map incorporates six fundamental causal patterns. The findings reveal that the experts' maps tend to under-represent the more intricate forms of causal structure associated with systems thinking. The authors argue that gaining an understanding of how individuals perceive sustainability provides valuable insights into the underlying foundations of normative debates and helps prevent unintended consequences that may arise from decisions focusing solely on isolated components of a system.

Okello et al. (2018) emphasizes the significance of high-quality seed potatoes in enhancing food security and reducing the volatility of food prices in developing nations. However, many smallholder farmers still rely on inferior seed obtained from local sources, which leads to lower yields. To explore the motivations and mental models related to the adoption of quality seed potatoes, the study employs means-end chain theory. The findings reveal that the decision to purchase and utilize quality seed is driven by farmers' deeply rooted life goals, with profit-making serving as a means to achieve these broader objectives. Additionally, the study observes variations in the structure of mental models based on the gender of the adopter and differences between adopters and non-adopters. The article concludes that farmers' mental models differ in structure and content based on the type of farmer, and discusses the policy implications stemming from these findings.

Henly-Shepard et al. (2015) discusses the utilization of participatory modelling as a means to promote social learning and facilitate community disaster planning, particularly in coastal island communities that face significant risks associated with climate change-induced natural hazards. It highlights the lack of available methods for communities to explicitly assess anticipated risks and potential adaptation strategies specific to their socio-ecological system. The article introduces a three-phase social learning framework aimed at enabling stakeholder-driven scenario-based modelling, with the purpose of informing community disaster planning in relation to the potential impacts of a tsunami. The participatory research was conducted in collaboration with a community disaster committee representing the North Shore communities of O'ahu, Hawai'i. Through a series of iterative participatory modelling workshops utilizing fuzzy-logic cognitive mapping, the committee actively engaged in representing, exploring, and questioning their beliefs regarding the natural hazards confronting their community. The modelling process enabled the committee to capture the dynamic nature of the community, simulate tsunami hazard scenarios to assess potential direct and indirect effects, and explicitly compare the trade-offs associated with different adaptation strategies. The changes observed in the committee's model representations over time signify a progression from single- to double- and triple-loop learning, indicating the occurrence of social learning across individual, institutional, and temporal levels.

Gray et al. (2014) discusses the difficulties encountered in identifying and defining climate change issues that necessitate adaptive measures in coastal management. The authors argue that in order for adaptation to advance, both local-scale "bottom-up" experts and decision-makers and national-scale "top-down" scientists and policymakers must detect climate signals and frame climate impact problems in a similar manner. To assess stakeholder perceptions of climate vulnerability at both individual and collective levels, the authors employ participatory modelling with coastal stakeholders, utilizing Fuzzy Cognitive Mapping (FCM). The study reveals that FCM not only provides insights into the current detection and framing of climate signals in coastal decision-making but also serves as a structured communication platform, allowing for the coherent integration of climate problems into future coastal management discussions as the adaptation process evolves.

Halbrendt et al. (2014) discusses the efforts to promote conservation agriculture in developing nations and the disparity between the perspectives of scientific experts and rural farming communities regarding the adoption of new farming practices and technologies. The study focuses on a case study conducted in Nepal, comparing the mental models of experts and farmers regarding the outcomes associated with technological advancements. The findings indicate that the implementation of conservation agriculture techniques should not be universally applied. Instead, it emphasizes the need for development practitioners to engage in mutual learning with local communities in order to benefit from locally contextualized knowledge. The article underscores the importance of considering a bottom-up perspective when introducing agricultural development programs and advocates for a participatory approach that involves the community and stakeholders.

Hoffman et al. (2014) emphasizes the significance of comprehending how farmers perceive and apply the concept of sustainability in their farm management decisions. The authors examine the "mental models" of sustainability held by farmers in three wine grape-growing regions of California, where local extension programs have concentrated on promoting sustainable agriculture. Mental models are represented as networks, with sustainability concepts as nodes, and links established when a farmer mentions two concepts in their definition of sustainability. The findings indicate that wine grape growers' mental models of sustainability exhibit a hierarchical structure, exhibit similarity across regions, and are strongly influenced by participation in extension programs and the adoption of sustainable farming practices. The article concludes that the concept of sustainability serves as a valuable framework for bridging the gap between knowledge and action. Moreover, it highlights a positive correlation between the sophistication of an individual's beliefs about sustainability, their engagement in local outreach programs, and their adoption of sustainable practices.

Nyaki et al. (2014) examines the application of participatory modelling to gain insights into the dynamics and underlying factors driving the bush meat trade in local communities in Tanzania. The study reveals that existing bush meat management policies, which are often externally developed, tend to overlook the variations in social and ecological conditions present in different communities. Through the use of participatory modelling, the study discovers that local communities possess a diverse range of drivers influencing bush meat trade dynamics. These factors include collaborative law enforcement, increasing human population, market demand, and cultural preferences. The study proposes that participatory modelling can foster alignment between community perspectives and conservation organizations, thereby facilitating the development of contextually appropriate policies that can be continuously monitored and adjusted based on new evidence.

Turner et al. (2014) examines the correlation between different land ethics held by agricultural stakeholders and recent changes in land use. The research delves into the historical evolution of land ethics in the United States and how they have influenced both land use policies and private lands. By collecting primary data through semi-structured interviews with farmers, ranchers, and influential stakeholders, the study utilizes stakeholder analysis methods to identify key factors considered in land use decisions, ascertain the priorities of each group regarding these factors, and define the relevant mental models that shape each group's perception of the land ethic concept. The findings reveal notable variations in the qualitative prioritization of land use factors and highlight distinct differences in land ethics among these stakeholder groups. The study concludes that shifts in stakeholder land ethics have played a role in recent land use changes. Lastly, the paper discusses how the existing agricultural policy sends conflicting signals regarding preferred land use practices and explores the potential consequences resulting from these divergent land ethics.

Gray et al. (2013) studied that the Mental Modeler is a software tool designed to facilitate user-driven model construction and encourage knowledge sharing among diverse stakeholder communities. By utilizing fuzzy-logic cognitive mapping (FCM), it enables users to develop and refine their models, with the goal of supporting adaptive management planning. The tool explicitly captures the mental models of stakeholders, thereby offering an opportunity to integrate various forms of knowledge into environmental decision-making processes. It allows users to formulate hypotheses, test them, and run scenarios to assess the potential outcomes of proposed policies. In recent years, there has been increasing recognition of the significance of involving stakeholders in environmental decision-making through modelling. Participatory modelling operates on a hierarchical scale, starting from individual "mental models" and extending to the systemic level. Mental models are internal constructs that shape how individuals interpret and structure their understanding of the external environment, playing a crucial role in decision-making. Modelling has long been acknowledged as a vital component of adaptive management initiatives. Walters identifies three key objectives of modelling in this context: (1) enhancing problem clarification and communication among scientists, managers, and stakeholders; (2) screening policies to identify options that are unlikely to be effective due to inadequate scale or impact type; and (3) identifying critical knowledge gaps that undermine the reliability of model predictions.

Gray et al. (2012) discusses the advantages and drawbacks of incorporating stakeholder knowledge into the management of natural resources. The integration of various knowledge systems is believed to enhance the resilience of social-ecological systems. Nevertheless, there are concerns regarding biases and the lack of concrete evidence supporting some of the claimed benefits. The article assesses the understanding of a model social-ecological system, specifically the summer flounder fishery in the mid-Atlantic, to compare the structural and functional aspects of stakeholder mental models. The research suggests that while there might be advantages to including diverse knowledge in resource decision-making, there are also associated costs.

Gray et al. (2012) explores the application of Fuzzy Cognitive Maps (FCM) as a tool for representing mental models and collective beliefs. FCM has been employed in various scientific fields to gain insights into the belief systems held by individuals and groups. The article delves into theoretical considerations regarding the use of FCM to represent mental models and their usefulness in comparing and characterizing the aggregated beliefs and knowledge within a community. It also discusses the theoretical underpinnings of concept mapping, cognitive mapping, mental models, and the concept of "expertise" in eliciting an individual's knowledge. The article further examines the use of FCMs obtained from non-traditional experts, aiming to analyse the structure of group knowledge to inform community decision-making and compare variations in knowledge across different groups. Ultimately, the article concludes by addressing how FCM can be employed to comprehend shared knowledge and highlights the trade-offs that should be considered when selecting FCM data collection techniques.

Reynolds et al. (2010) highlights the significance of assessing the public's understanding of climate change to enhance effective communication and decision-making. The authors conducted exploratory studies and mental model interviews to gain insights into the public's comprehension of climate change. The findings reveal that many individuals believed global warming to be both concerning and probable, often confusing stratospheric ozone depletion with the greenhouse effect and mistaking weather for climate. Additionally, respondents frequently attributed primary causes of global warming to automobile usage, heat and emissions from industrial processes, aerosol spray cans, and pollution in general. The article underscores the importance of taking these beliefs into account when designing risk communications or presenting climate-related policies to the public. Furthermore, it discusses the authors' interview protocol and highlights its distinctions from other studies on the subject matter.

4

Social Ecology

The research setting provides specific details about the location where a study is conducted. This study area typically encompasses a particular geographical region, such as a state, district, block, or gram panchayat, chosen according to the research requirements. The researcher must have a thorough understanding of the locality, including its communication methods, in order to effectively access all parts of the region for data collection.

In social science research, a clear comprehension of the characteristics, attitudes, and behaviours of the individuals within the study area is essential. Without this understanding, accurately conceptualizing, perceiving, and interpreting data becomes challenging. To obtain a comprehensive understanding of the implications and behavioural intricacies of the local people representing a broader community in the referenced study area, researchers must have a solid grasp of these factors. Therefore, this chapter critically analyses the socio-demographic background of the residents in a rural setting.

The term "research setting" denotes the context in which research inputs and elements are absorbed, interacted with, and mutually contribute to the system's overall performance.

The research setting is crucial because it shapes and influences the interactions of various factors and components. Therefore, conducting a study on how farmers perceive persuasive issues and other topics requires a local context that takes into account the unique natural environment, demographics, crop ecology, institutional framework, and socio-cultural factors. For example, when writing a paper about the social behaviour of chimpanzees, it is important to include details about the research setting. This may include information such as where the chimpanzees were studied (in the wild or in captivity), the number of chimpanzees that were observed, whether they were part of the same social group, the geographic location, the time of the study, the season/weather conditions, the availability of resources such as food, water, and shelter, and any environmental factors that were present.

There are two types of research settings: macro and micro. The macro research setting encompasses the entire state, while the micro research setting focuses on specific districts, blocks, or village peripheries. The aim of this approach is to understand the study environment from broader perspectives using the state as a reference frame and the district and block profiles as units of in-depth analysis.

This study was conducted in JalkarBhomra, Kastadanga and Subuddipur villages of Gram Panchayat-Kastadanga-1 of Haringhata block in Nadia district, West Bengal. The state and district, as well as the block and villages in particular, are briefly described in this chapter.

4.1. Area of Study

This study focuses on a region situated in the eastern part of India, namely the state of West Bengal. This state has distinct social and ecological characteristics that impact the behaviours and living standards of its people. The study is specifically focused on the Haringhata block in the Nadia district which includes the villages of JalkarBhomra, Kastadanga and Subuddipur of the Gram Panchayat-Kastadanga-1.

4. 2. Profile of the State

West Bengal is a state in eastern India that shares its northern boundary with Sikkim and parts of Nepal and Bhutan, eastern boundary with Bangladesh, western boundary with Indian states of Orissa, Bihar, and Jharkhand, and to the southern part of the state lies the Bay of Bengal. It extends between 85°48'20"E to 89°53'04"E longitude and 27°13'15"N to 21°25'24"N latitude. The state capital is Kolkata, which is also the third-largest metropolis and seventh largest city by population in India. It covers an area of 88,752 sq. km and is divided into 20 administrative districts. West Bengal includes the Darjeeling Himalayan hill region, the Ganges delta, the Rarh region, and the coastal Sundarbans. The state's main ethnic group is the Bengalis, with Bengali Hindus forming the demographic majority.

Post-Indian independence, West Bengal's economy was based on agricultural production and small and medium-sized enterprises. The economy of West Bengal is the sixth-largest state economy in India with a gross state domestic product (GSDP) of INR 12.54 lakh crore (US$180 billion), and has the country's 23rd-highest GSDP per capita of INR115,748 (US$1,600). West Bengal has two World Heritage sites and ranks as the seventh-most visited tourist destination in India.

The state of West Bengal was formed on November 1, 1956 and is presently bifurcated into 23 districts under five divisions. These divisions are the Burdwan division, Jalpaiguri division, Presidency division, Medinipur division, and Malda division. The districts are Alipurduar, Bankura, Birbhum, Cooch Behar, Darjeeling, Dakshin Dinajpur, Hooghly, Howrah, Jalpaiguri, Jhargram, Kolkata, Kalimpong, Malda, Murshidabad, Nadia, North 24 Parganas, Paschim Bardhaman, Paschim Medinipur, Purba Bardhaman, Purba Medinipur, Purulia, South 24 Parganas, and Uttar Dinajpur. The state stretches over an area of 88,752 sq. km. with 66 sub-divisions, 121 municipalities, 6 municipal corporations, 341 development blocks, 40,782 Mouzas, 3,349 Gram Panchayats, and 37,469 inhabited villages (Patra, 2022 and Chandra, 2022).

Map 1: Map of West Bengal (Source - Apoorv, 2011)

4.2.1 General Information of the State

Table 1: General Information of the State

Population	
Male	46,809,027
Female	44,467,088
Sex Ratio	947
Rank	4th
Urban %	31.87
Rural %	68.13
Density	1,000/km2 (2,700/sq. mi)
Growth	13.84%
Highest Density	Kolkata (24,306)
% of total	1,01,12,599
Boys	51,87,264
Girls	49,25,335
Sex Ratio	950
Literacy	
Literate (Total)	6,15,38,281
Male	3,38,18,810
Female	2,77,19,471
Literacy Rate	77.08%
Male	82.67%
Female	71.16%

(Source - Department of Information and Cultural Affairs (I&CA), Government of West Bengal)

4.2.2 Geography and Climate of the State

West Bengal is a state located on the eastern bottleneck of India and stretches from the Himalayas in the north to the Bay of Bengal in the south. The state has a total area of 88,752 square kilometres (34,267 sq. mi). The Darjeeling Himalayan hill region in the northern part of the state is a part of the eastern Himalayas mountain range, and Sandakfu, the highest peak in the state, is located there at 3,636 m (11,929 ft.). The narrow Terai region separates the hills from the North Bengal plains, which in turn transitions into the Ganges delta towards the south. The Rarh region intervenes between the Ganges delta in the east and the western plateau and highlands. A small coastal region is in the extreme south, while the Sundarbans mangrove forests form a geographical landmark at the Ganges delta. The Ganges River is the main river in West Bengal and divides into two branches. One branch enters Bangladesh as the Padma, while the other flows through West Bengal as the Bhagirathi River and Hooghly River. The Teesta, Torsa, Jaldhaka, and Mahananda rivers are in the northern hilly region. The western plateau region has rivers like the Damodar, Ajay, and Kangsabati. The Ganges delta and the Sundarbans area have numerous rivers and creeks. Damodar, another tributary of the Ganges, has several dams under the Damodar Valley Project. Arsenic contamination of groundwater affects at least nine districts in the state, with an estimated 1.04 crore people afflicted by arsenic poisoning as of 2017.

The climate of West Bengal varies from tropical savanna in the southern portions to humid subtropical in the north. The northern part of the state is characterized by the sub-tropical monsoon, mild to dry winter, and hot summer, while the southern part experiences a tropical savanna with hot and dry winter. The Darjeeling district in the north has a climatic type of tropical upland, mild and dry winter, and short warm summer. The main seasons are the summer, the rainy season, a short autumn, and the winter. The summer in the delta region is noted for excessive humidity, while the western highlands experience a dry summer like northern India. The highest daytime temperatures range from 38°C (100°F) to 45°C (113°F), with the hottest months being May and April. In early summer, brief squalls and thunderstorms known as Kalbaisakhi, or Nor'westers, often occur. West Bengal receives the Bay of Bengal branch of the Indian Ocean monsoon that moves in a southeast to northwest direction, bringing rain to the whole state from June to September. Heavy rainfall of above 250 centimetres (98 in) is observed in the Darjeeling, Jalpaiguri, and Cooch Behar district. Winter (December–January) is mild over the plains with average minimum temperatures of 15°C (59°F), and a cold and dry northern wind blows, substantially lowering the humidity level. The Darjeeling Himalayan Hill region experiences a harsh winter, with occasional snowfall. Relative humidity is generally high during the period from July to September, and it is about 80% in June, rising to about 83%-85% in July, August, and September in the morning. The diurnal variation of relative humidity is least during the monsoon season. The relative humidity is lowest during the summer afternoons when it becomes about 38% to 50% at the plain.

4.2.3 Agriculture of the State

West Bengal is primarily an agrarian state, accounting for only 2.7% of India's land area but supporting almost 8% of the country's population. The state has 71.23 lakh farm families, with 96% being small and marginal farmers who hold an average land size of only 0.77 ha. Despite this, West Bengal is rich in natural resources and has diverse agro-climatic conditions, enabling the cultivation of a wide range of crops. The state is the leading producer of paddy and vegetables in India, and its net cropped area comprises 68% of the total land area and 92% of the arable land. However, West Bengal faces frequent natural calamities such as floods, cyclones, and hailstorms due to its location in the humid tropic region close to the Bay of Bengal.

Although West Bengal has a surplus production of rice, vegetables, and potatoes, it faces a significant gap between the production and requirement of pulses, oilseeds, and maize. Moreover, several challenges such as soil health deterioration due to the misuse of chemical fertilizers, inadequate farm mechanization, and an unorganized marketing structure hinder agricultural growth in the state.

More than 72% of the rural population of West Bengal earns their livelihood from agriculture, and the total cultivable area of the state contributes to 65.65% of the total land area. The state's net sown area is 5.234 mha, and its fertile alluvial soil, which predominates over a large area, makes it highly productive. The state also benefits from both tropical and temperate climatic conditions, enabling diversification in agriculture. Due to its high potential in the agriculture sector, the central government included West Bengal as one of the seven states in the "Bringing Green Revolution to Eastern India (BGREI)" Programme of 2011.

Table 2: Agricultural statistics of West Bengal (Source: Chandra, 2022)

1.	Total reported area	8.68 mha
2.	Total cultivable area	5.65 mha
3.	Total forest area	1.17 mha
4.	Cultivable wasteland	16940 ha
5.	Permanent pasture and other grazing land	2370 ha
6.	Gross cultivated area	9.6 mha
7.	Net cultivated area	5.2 mha

Table 3: Land use pattern of West Bengal (Source: Patra, 2022)

Geographical Area (mha)	8.86
Net Cropped Area (mha)	5.46
Gross Cropped Area (mha)	9.23
Irrigated Area (mha)	4.67
Cropping Intensity (%)	165.34
Forest Area (%)	13.72
Net Agricultural Use (%)	18.66
Barren and Uncultivable (%)	0.35
Permanent pastures and Grazing (%)	0.09
Trees and Groves (%)	0.88
Cultivable wasteland (%)	0.53
Fallow land (%)	0.34

The cropping intensity in West Bengal is much higher than the national average of 142%, with an average of 185%. However, the state's fertilizer consumption ratio of N, P, and K is much lower than the recommended dose of 4:2:1. This clearly indicates an imbalance in nutrient supply to the plants, which ultimately leads to lower production and productivity. This disparity in fertilizer consumption also suggests that the soils in West Bengal are deficient in nutrients. Moreover, the farmers' reliance on chemical fertilizers has led to increased soil acidity in many parts of the state. In addition, the Sundarbans region in the southern part of the state faces issues with soil salinity due to its exposure to the brackish water of the Bay of Bengal.

West Bengal's major crops include food crops such as paddy, wheat, maize, and pulses like lentil, mung, arhar, gram, and lathyrus. Other crops include potato, vegetables like brinjal, tomato, okra, cabbage, cauliflower, cucurbits, onion, and other leafy vegetables, and oilseeds like rapeseed and mustard, sesame, linseed, and others. The state's most important cash crop is jute, while it is also known for its world-famous Darjeeling tea. The districts of Hooghly and Burdwan are considered to be the most productive for agriculture and together form the "Rice Bowl" of the state. Additionally, other agriculturally rich districts include Nadia, Medinipore, both 24 Paraganas, and Coochbehar. The state is also rich in horticulture, producing fruits such as mango, banana, pineapple, papaya, guava, jackfruit, litchi, citrus, and sapota. Furthermore, the state is home to various livestock, including cattle, buffaloes, sheep, pigs, and goats in substantial numbers.

4.2.4 Description of Agro-ecological Zone

West Bengal is divided into six agro-ecological zones, each with unique and distinct features. These zones include the Hill zone, Terai zone, Old alluvial zone, New alluvial zone, Red and lateritic zone, and Coastal and saline zone.

The present study focuses on the new alluvial zone only. The New alluvial zone covers an area of 15.304 lakh hectares, which is 55.7% of the state's total geographical area and encompasses districts such as Nadia, Howrah, Hooghly, North 24 Parganas, Kolkata, parts of Bardhaman, and parts of South 24 Parganas. The region experiences an annual precipitation of 1350-1450mm and an average temperature ranging between 35 degrees Celsius to 15 degrees Celsius. The soil texture is mainly loamy to clay in nature and is highly fertile due to the alluvium of the Ganges. The region grows a variety of agricultural crops such as rice, wheat, mustard, potato, seasonal vegetables, and cash crops such as jute and Mesta.

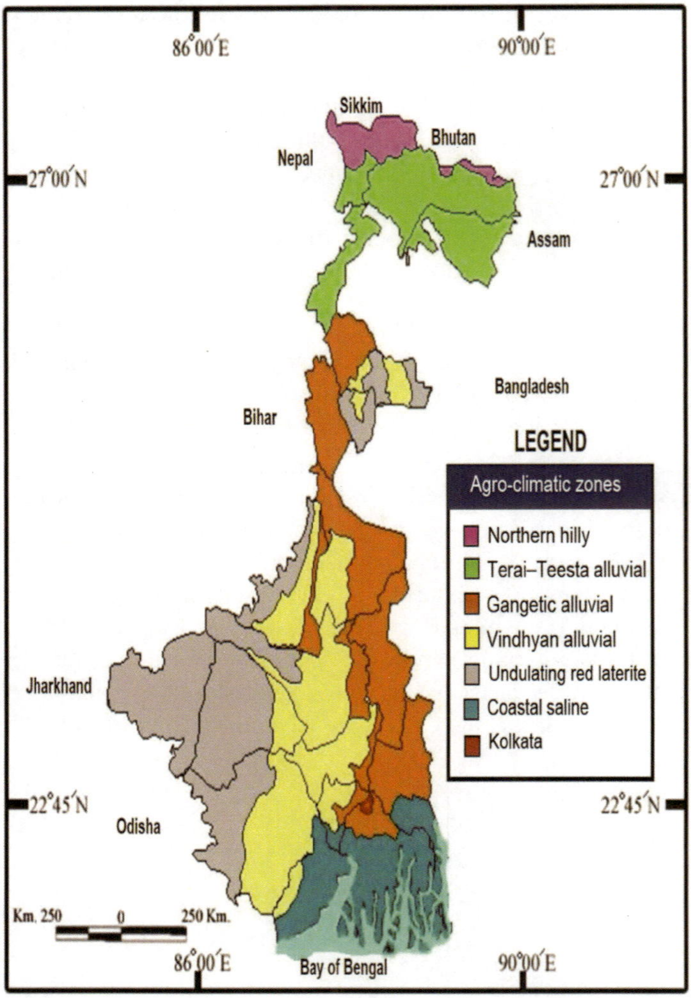

Map 2: Agro-ecological map of West Bengal (Source: Mandal et al., 2022)

4.3. Profile of the District

Nadia district is located in the heart of the Bengal delta, surrounded by the Ganga's arms, namely the Bhagirathi on the west and Mathabhanga on the north. It is situated in the alluvial plain of the Ganga and its tributaries. Unlike other districts in West Bengal, Nadia district is not named after its headquarters station, Krishnagar, but after Nadia or Nabadwip, which is revered for being the birthplace of Lord Shri Chaitanya Mahaprabhu on February 18, 1486. The British district of Nadia was established in 1787, and the present district of Nadia was formed after the partition by Notification No.545-GA dated February 23, 1948. Nadia district shares its borders with Murshidabad to the north and northwest, the Republic of

Bangladesh to the southeast and east, and North 24-Parganas to the south and southeast. The district has an irregular shape, extending from north to south. With an elevation of approximately 46 feet above sea level, the district is divided into two parts by the Tropic of Cancer. The climate in Nadia is characterized by a hot summer, high humidity throughout the year, and well-distributed rainfall during the southwest monsoon season. The cold season lasts from the end of November to the end of February. Krishnanagar, located on the banks of the river Jalangi, serves as the district headquarters and is the most significant town in Nadia district. The Nadia Raj Palace (Rajbari) in Krishnanagar is a popular tourist attraction, although the grandeur of the past has faded, leaving behind a dilapidated structure with exquisite carvings on its inner walls. Krishnanagar is renowned for its world-class clay models and terracotta works. It is also known for its delicious sweets, particularly Sarpuria and Sarbhaja. Kalyani, another important town and sub-divisional headquarters in Nadia, was identified and named by Dr. B. C. Roy, the former Chief Minister of West Bengal. Located about 50 km away from Kolkata, Kalyani serves as an alternative to the city. The sub-division houses two universities, two engineering colleges, and a recently established medical college.

Nadia district is linear in shape, stretching between 22º53″ and 24º11″ North latitude and 88º09″ and 88º48″ East longitude. It is situated at an elevation of approximately 46 feet above sea level, divided into two parts by the Tropic of Cancer. The district shares its geographical boundaries with Bangladesh to the east, Bardhaman and Hooghly districts to the west, Murshidabad district to the north and northwest, and North 24 Parganas to the south and southeast. The district is easily accessible by rail, with major railway stations such as NabadwipDham and Ranaghat providing regular train services to Kolkata, Howrah, NJP, and Guwahati. The International Rail Link connecting India and Bangladesh passes through Nadia district, with Gede being the last railway station on the Indian side. Nadia district is well-connected by road, with National Highway-34 passing through its northern and eastern parts. It can be accessed by road from various parts of the country. The nearest airport, Kolkata, is located between 100 to 160 kilometres from different parts of the district. The district headquarters, Krishnagar, is approximately 110 kilometres from Kolkata and can be easily reached via regular bus services from Kolkata and North Bengal. The district also has waterway connectivity through rivers such as Bhagirathi, Jalangi, Bhairav, Churni, Mathabhanga, and Ichamati.

4.3.1 General Information of the District

Table 4: General Information of the District

Total population	5,167,600
Male population	2,653,768
Female population	2,513,832
Population Growth	12.22%
Sex ratio (per 1000)	947
Population density	1300/ km^2

Total Literacy Rate	74.97%
Literacy Rate (Male)	78.75%
Literacy Rate (Female)	70.98%
No. of Sub-division	5
No. of Development Block	17
No. of Gram Panchayat	187
No. of inhabited Village	2639
Area (Sq. Km)	3,927

(*Source*: Office of the Register General and Census Commissioner 2011 Ministry of Home Affairs, GOI)

4.3.2 Climate of the District

The climate in Nadia district is similar to that of other areas in West Bengal, except for the colder Northern region. The cold season begins in mid-November and lasts until the end of February. The main characteristics of the climate are hot summers, high humidity, and evenly distributed rainfall during the South-West Monsoon season. In March and April, the weather becomes increasingly hot, with high temperatures during the day. The South-West Monsoon typically begins in early June and continues until late September.

Table 5: The maximum and minimum temperature of the district for the preceding

Year	Maximum temperature (°C)	Minimum temperature (°C)
2007	39	8
2008	37	9
2009	41	8
2010	41	7
2011	37	5

(*Source*: District Statistical Hand Book, Nadia, 2010-11)

Table 6: The status of Annual Rainfall during the period of 2007-2011

Year	Annual Rainfall in mm
2007	1612
2008	1436
2009	1209
2010	940
2011	1488

(*Source*: District Statistical Hand Book, Nadia, 2010-11)

4.3.3 Agriculture of the District

Throughout the year, Nadia district maintains its lush greenery with seasonal crops, and almost all significant crops are cultivated successfully in the region. The district's economy primarily relies on agriculture.

Yield rates of some Selected Crops iu the district of Nadia aud West Bengal

(Kilogram per hectare)										
Crops	**2009-10**		**2010-11**		**2011-12**		**2012-13**		**2013-14**	
	District	**West Bengal**	**District**	**West Bengal**	**District**	**West Bengal**	**District**	**West Bengal (P)**	**District**	**West Bengal**
(D	(2)	(3)	(4)	(5)	(6)	(7)	(8)	(9)	(10)	(11)
Rice	2784	2547	2759	2708	2953	2688	2870	2745	3072	2789
Wheat	2549	2680	2536	2760	3078	2765	3320	2786	3366	2799
Gram	861	1110	853	1069	1095	1049	913	1176	1074	1175
Jute	2898	2573	2466	2576	2502	2572	2554	2569	3128	2788
Rapeseed & Mustard	1100	850	1028	1021	949	908	1193	1062	1177	1066
Potato	30676	35768	27301	32831	30959	25641	31905	29869	30590	21955
Tea	-	1940(E)	-	1984(E)	-	1918(R)	-	2046 (R)	-	2222

Sources: 1) Directorate of Agriculture, Govt, of W.B

2) BAE & S., Govt, of W.B.

3) Tea Board

4.3.3.1 Soil and Cropping Pattern of the District

Nadia district falls under the new alluvial agro-climatic zone of WB, which is conducive to the successful cultivation of various important crops. The district's economy is heavily reliant on agriculture. The soil is particularly suitable for growing paddy, and in many areas of the district, farmers also cultivate wheat, pulses, and jute. Additionally, the low-lying region known as Kalantar in the district is well-suited for the cultivation of aman rice.

4.3.3.2 Land Use Pattern of the District

Nadia district consists of highly fertile land that is well-suited for cultivating various agricultural crops. The district focuses on cultivating important crops such as paddy, wheat, jute, and pulses. Additionally, high-quality vegetables are also produced in the district. The table below illustrates the land use pattern in the district.

Classification of Land Utilisation Statistics in the district of Nadia

(Thousand hectares)

Year	Reporting Area	Forest Area	Area under Non-agricultural use	Barren & unculturable land	Permanent pastures & other grazing land	Land under misc. tree groves not included in 'Net area sown'	Culturable waste land	Fallow land other than Current fallow	Current fallow	Net area sown
' (D	(2)	(3)	(4)	(5)	(6)	(7)	(8)	(9)	(10)	(11)
2009-10	390.66	1.22	88.94	0.24	0.05	3.91	0.79	0.37	4.40	29074
2010-11	390.66	1.22	90.22	0.06	0.06	3.73	0.63	0.11	4.18	290.45
2011-12	390.66	1.22	88.45	0.03	0.01	2.58	0.58	0.59	4.26	292.94
2012-13	390.66	1.22	89.02	0.03	-	2.49	0.40	0.52	3.43	293.55
2013-14	390.66	1.22	89.15	0.03	-	2.28	0.31	0.43	2.94	294.30

Source : Directorate of Agriculture(Evaluation), Govt, of W.B.

Distribution of Operational Holdings over size-classes in the district of Nadia

(Area in Hectare)

SIZE - CLASS													
Year	**Marginal**		**Small**		**Semi-mdium**		**Medium**		**Large**		**Total**		**Average Size of holdings (hectare)**
	No. of holdings	**Area of holdings**	**No. of holdings**	**Area of holdings**	**No. of holdings**	**Area of holdings**	**No. of holdings**	**Area of holdings**	**No. of holdings**	**Area of holdings**	**No. of holdings**	**Area of holdings**	
(1)	(2)	(3)	(4)	(5)	(6)	(7)	(8)	(9)	(10)	(11)	(12)	(13)	(14)
2005-06	329150	182763	75752	128679	12443	33910	880	4822	13	415	418238	350589	0.84
2010-11	338475	188532	73274	123849	11937	34547	500	2534	20	526	424206	349988	0.83

Small: 1 0 hectare and above but less than 2.0 hectares

Semi-medium: 2 0 hectares and above but less than 4 0 hectares

Medium: 4.0 hectares and above but less than 10.0 hectares

Large: 10.0 hectares and above. It includes mostly institutional holdings

1Acre: 0.404686 hectare

Source : Agricultural Census, West Bengal

4.3.3.3 Irrigation

The primary source of irrigation in the district is rivers, while lift irrigation and tube-wells are also utilized to water the cultivated fields. The table below displays the different sources of irrigation.

Sources of Irrigation in the district of Nadia

(Number)

Year	Tank	HDTW	MDTW	LDTW	STW*	RLI	ODW	Others
(1)	(2)	(3)	(4)	(5)	(6)	(7)	(8)	(9)
2009-10	-	609	31	216	502	293	-	1
2010-11	-	612	31	228	588	293	-	1
2011-12	-	637	35	264	567	319	-	-
2012-13	-	565	31	24	458	319	-	-
2013-14	-	580	34	24	458	311	-	-

Note : HDTW = High capacity Deep Tubewell
MDTW = Middle capacity Deep Tubewell
LDTW = Low capacity Deep Tubewell
STW = Shallow Tubewell
RLI = River Lift Irrigation
ODW = Open Dug Well
Sources: 1) Asstt. Engineer, Agricultural Meeh. Nadia
2) Asstt. Engineer, Agri. Irrigation, Nadia
3) Water Investigation & Development
* Only Govt.
Department, Govt of W B.

Area Irrigated by different sources iu the district of Nadia

(Thousand hectares)

Year	Area irrigated by									
	Govt. Canal	Tank	HDTW	MDTW	LDTW	STW*	RLI	ODW	Others	Total
(1)	(2)	(3)	(4)	(5)	(6)	(7)	(8)	(9)	(10)	(11)
2009-10	-	-	22.03	0.62	1.31	2.87	11.70	-	0.01	38.54
2010-11	-	-	22.09	0.60	1.30	2.18	11.70	-	0.01	37.88
2011-12	-	-	21.85	0.60	1.30	2.34	9.40	-	-	35.49
2012-13	-	-	11.79	0.44	0.06	1.03	10.68	-	-	24.00
2013-14	-	-	14.96	0.40	0.04	0.75	10.61	-	-	26.77

Note : HDTW = High capacity Deep Tubewell MDTW = Middle capacity Deep Tubewell LDTW = Low capacity Deep Tubewell	STW = Shallow Tubewell RLI = River Lift Irrigation ODW = Open Dug Well	Sources: 1) Irrigation and Waterways Directorate, Govt, of W.B 2) Asstt. Engineer, (Agri Meeh.) & (Agri. Irri), Nadia
* Only Govt.		

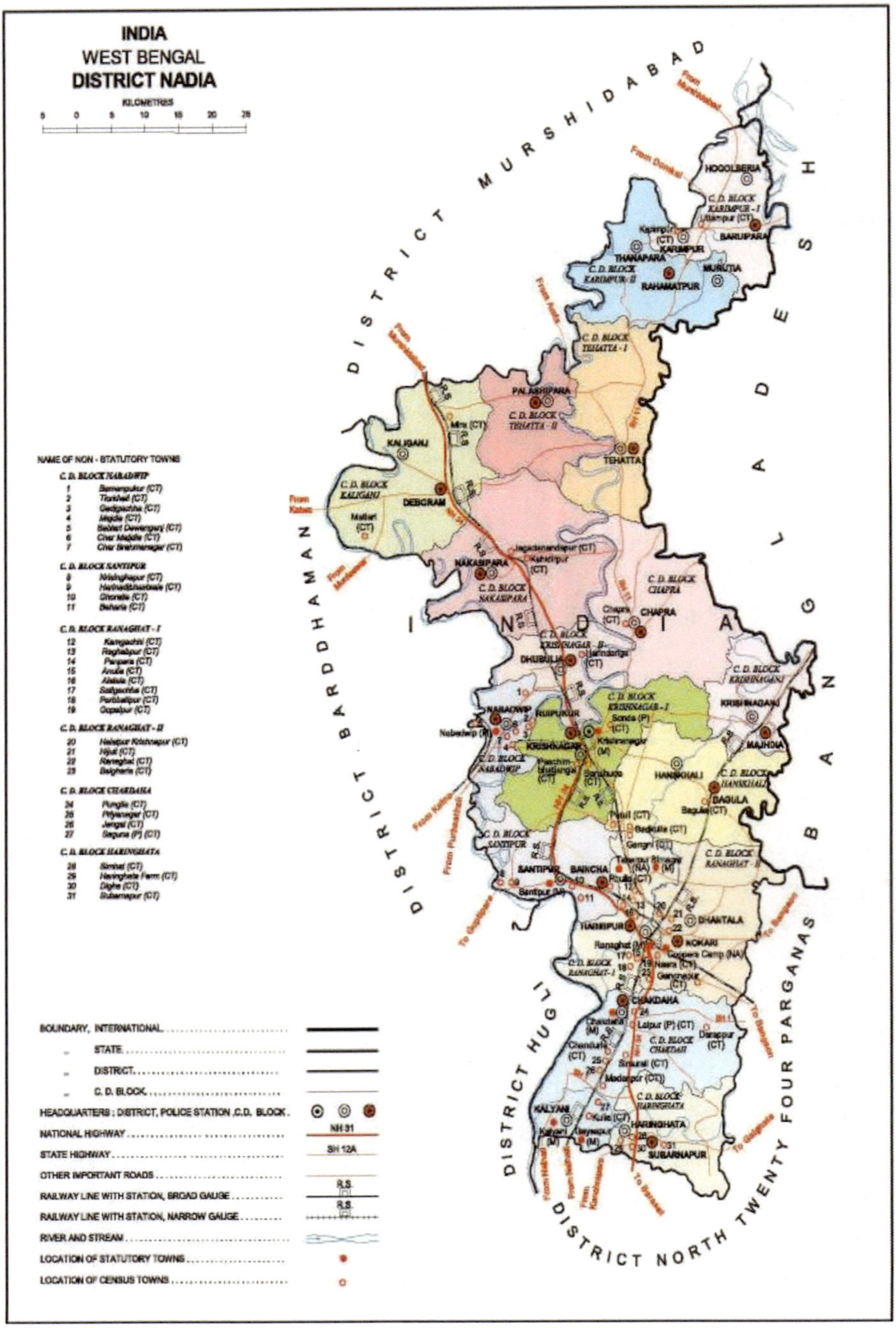

Map 3: Map of Nadia District, West Bengal (Source: Census Handbook, Nadia, 2011)

4.4. Profile of the Block

Haringhata Development Block situated in the Kalyani subdivision of the Nadia district in the Indian state of West Bengal, has got its present status after bifurcation and creation of Haringhata Municipality with two gram panchayats viz. Haringhata-1 and Haringhata-2 since 2nd January, 2014. Presently the block consists of 8 gram panchayats (Viz. Birahi-1, Birahi-2, Fatepur, Mollabellia, Kastadanga-1, Kastadanga-2, Nagarukhara-1 and Nagarukhara-2). Haringhata block is the southernmost block of the Nadia district and is surrounded by North 24 Parganas district from 3 sides. Positioned at coordinates 22°95′N and 88°57′E, Haringhata has an average elevation of 10 meters above sea level. The block is blessed with fertile agricultural land and thereby creating a progressive farming community. The block has also become a vibrant educational hub with presence of 4 universities (BCKV, MAKAUT, IISER-K and WBUA&FS). Presence of AIIMS Kalyani at the border of Haringhata police station jurisdiction has also changed the health scenario of the locality. The predominant language spoken in this region is Bengali.

4.4.1 General Information of the Block

Table 7: General Information of the Block

Block area	170.32 sq. km
No. of Gram Panchayat	8
No. of Gram Sabha	129
No. of Mouza	87
No. of Police Station	1
Elevation	13 m (43 ft.)
Total Population (2011)	2,31,068
Population Density	1,400/km^2
Languages	Bengali, English
Total literates	1,71,144 (82.15%)

(*Source*: Block Development Office, Haringhata)

4.4.2 Geography of the Block

Haringhata CD Block is surrounded by Chakdah Block to the north and west, and Bangaon and Gaighata Blocks in the east located in the North 24 Parganas district. To the south, it is bordered by Amdanga, Habra I, and Habra II Blocks, also in the North 24 Parganas district. The total area of Haringhata Block is 170.32 square kilometres, and it comprises of 1 panchayat samiti, 10 gram panchayats, 176 gram sansads (village councils), 87 mouzas, and 82 inhabited villages. The block is served by a police station, and its headquarters are located in Subarnapur, which is situated 56 kilometres away from Krishnanagar, the district headquarters.

4.4.3 Demographics of the Block

4.4.3.1 Population

According to the 2011 Census of India, the population of Haringhata Block was 231,068, with 207,459 residing in rural areas and 23,609 in urban areas. The male population accounted for 118,709 (51%), while the female population was 112,359 (49%). Scheduled Castes accounted for 73,755 individuals (31.92%), and Scheduled Tribes accounted for 11,218 individuals (4.85%). In the 2001 census, the total population of Haringhata Block was 207,867, with 108,011 males and 99,856 females. The block experienced a population growth rate of 16.92% during the 1991-2001 decade, while the district's growth rate was 19.51%. The decadal growth rate for West Bengal was 17.84%. Haringhata Block includes four census towns: Haringhata (3,989), Simhat (6,949), Subarnapur (5,759), and Digha (6,916) (based on the 2011 census). Additionally, there are several large villages in Haringhata Block with populations exceeding 4,000, including Chandirampur (4,740), Narayanpur (4,548), Uttar Datta Para (5,430), Uttar Brahmapur (5,371), Subuddipur (6,051), Sekendarpur (7,917), Kathdanga (7,417), Mollabalia (5,889), Ganguria (4,291), Fatepur (5,093), Barasat (4,046), Ukrah (13,548), Goaldob (4,883), and Dighalgram (5,390) (according to the 2011 census).

4.4.3.2 Literacy

According to the 2011 census, Haringhata CD Block had a total of 171,144 literate individuals, which accounted for 82.15% of the population aged 6 years and above. Among them, 93,313 were males, representing 87.11% of the male population aged 6 years and above, while 77,831 were females, accounting for 76.90% of the female population aged 6 years and above. The gender disparity, measured by the difference in literacy rates between males and females, was 10.21%.

4.4.3.3 Language and Religion

The local language spoken in these areas is Bengali. According to the 2011 census, Hindus accounted for 164,263 individuals, comprising 71.09% of the population in Haringhata Block. Muslims accounted for 65,148 individuals, representing 28.19% of the population. Christians accounted for 460 individuals, making up 0.20% of the population. The remaining 1,197 individuals belonged to other religious groups, forming 0.52% of the population.

In the 2001 census, Hindus numbered 151,379, accounting for 72.79% of the population in Haringhata CD Block. Muslims numbered 55,885, making up 26.97% of the population. In the 1991 census, Hindus numbered 130,125, comprising 73.19% of the population in Haringhata CD Block. Muslims numbered 47,284, representing 26.60% of the population.

4.4.4 Economy

4.4.4.1 Livelihood

In 2011, Haringhata CD Block had a distribution of workers across various categories. Among the total workers, cultivators accounted for 18.42%, agricultural labourers for 35.60%, household industry workers for 3.79%, and other workers for 42.19%. The southern part of Nadia district, stretching from Krishnanagar I to Chakdah and Haringhata, consists of certain urban pockets that specialize in either manufacturing or service-related economic activities. This region has shown a relatively higher concentration of population, but the urban population has generally remained stagnant. Despite this, Nadia district still has a significant portion of its population residing in rural areas.

4.4.4.2 Infrastructure

There are a total of 82 populated villages in Haringhata Block. All of these villages, constituting 100%, have access to power supply and drinking water supply. Out of the 82 villages, 13 of them (15.85%) have post offices. Furthermore, all 82 villages (100%) have telecommunication services, including landlines, public call offices, and mobile phones. 64 villages (78.05%) have well-built roads for convenient access, while 30 villages (38.59%) have transportation facilities such as bus services, rail connections, or navigable waterways. Additionally, 6 villages (7.32%) have agricultural credit societies, and 8 villages (9.76%) have banks.

However, it is important to note that although all villages in Nadia district had power supply in 2011, a survey conducted in 2007-08 revealed that less than 50% of households had electricity connections. In rural areas of the country, tube wells were historically relied upon for safe drinking water. However, due to the public concern over arsenic contamination in groundwater, this is no longer considered a viable option. Unfortunately, piped water supply remains a distant aspiration. In 2007-08, the availability of piped drinking water in Nadia district was as low as 8.6%, significantly below the state average of around 20%.

4.4.4.3 Agriculture

Although the Bargadari Act of 1950 acknowledged the rights of bargadars to receive a greater portion of crops from the land they cultivated, it was not fully implemented. Wealthy landlords still retained large areas of land that exceeded the prescribed limit for land ownership. Significant land reforms were carried out in West Bengal from 1977 onwards. Excess land beyond the permitted ceiling was acquired and distributed among the peasants. This resulted in a transformation of the land ownership pattern. In the year 2013-14, the composition of individuals engaged in agriculture in the Haringhata Block was as follows: bargadars accounted for 2.44%, Patta holders (document holders) constituted 7.95%, small farmers (owning land between 1 and 2 hectares) represented 6.18%, marginal farmers (owning land up to 1 hectare) comprised 33.53%, and agricultural labourers made up 49.09%. Due to the significant proportion of agricultural labourers, the actual wage in the agricultural sector has been a matter of concern. During the same period, the Haringhata CD Block contained 207 fertilizer depots, 40 seed stores, and 44 fair price shops.

In the year 2013-14, the Haringhata CD Block produced various crops as follows: 72,866 tonnes of Aman paddy, the primary winter crop, cultivated on 24,942 hectares; 20,348 tonnes of Boro paddy (spring crop) grown on 5,720 hectares; 5,413 tonnes of Aus paddy (summer crop) harvested from 2,013 hectares; 4,317 tonnes of wheat produced from 1,315 hectares; 42,799 tonnes of jute harvested from 2,383 hectares; and 16,616 tonnes of potatoes obtained from 536 hectares. Additionally, pulses and oilseeds were also cultivated. The total irrigated area in the Haringhata Block in 2013-14 amounted to 967 hectares, with 89 hectares irrigated by river lift irrigation, 830 hectares by deep tube wells, and 48 hectares by shallow tube wells.

4.4.4.4 Haringhata Dairy Farm

To meet the demand for milk among the urban population of West Bengal, the Directorate of Dairy Development, operating under the Animal Resources Development Department of the West Bengal government, established the first dairy plant in Haringhata in 1950. Initially, the plant produced 1.5 thousand litres of bottled milk per day during the 1950-51 period. Over time, the production increased with the expansion of infrastructure. The dairy operation expanded to include two dairy plants, seven milk colony units, and 19 Milk Collection cum Chilling stations known as the "Greater Calcutta Milk Supply Scheme." By the 1980s, the daily milk production had reached 210 thousand litres. (The Directorate of Dairy Development operates under the Animal Resources Development Department, Government of West Bengal).

4.4.4.5 Banking

During the fiscal year 2013-14, the Haringhata Block housed the branches of 12 commercial banks and 2 Gramin banks.

4.4.5. Education

In the year 2013-14, the Haringhata Block had a total of 120 primary schools with an enrolment of 9,251 students. Additionally, there were 11 middle schools with 866 students, 3 high schools with 1,819 students, and 18 higher secondary schools with 23,167 students. The block also had one general college with 3,233 students, two technical/professional institutions with 1,880 students, and 453 institutions catering to special and non-normal education with 10,874 students. Among the 82 inhabited villages in Haringhata Block, four villages did not have any schools, while 21 villages had more than one primary school. Furthermore, 16 villages had at least one primary school and one middle school, and 14 villages had one middle school and one secondary school.

Map 4: Map of Haringhata CD Block (Source: Office of the BDO, Haringhata)

4.5. Profile of the Kastadanga village

Kastadanga is a village situated in the Haringhata Block of Nadia District in the state of West Bengal, India. The village of Kastadanga is part of the Kastadanga-I gram panchayat and belongs to the Presidency Administrative Division. It is located approximately 59 kilometres south to the district headquarter Krishnanagar, Nadia, and 7 kilometres from Haringhata. The state capital, Kolkata, is about 60 kilometres away. The PIN code of Kastadanga-I is 741257, and the postal head office is Nagarukhra. Kastadanga-I gram panchayat area is surrounded by Habra-I Block to the south, Chakdah Block to the west, and Gaighata Block to the east. This place is located on the border of Nadia District and North 24 Parganas District. The local language spoken here is Bengali. The total population of the village is approximately 18,019, with around 9,817 males and 8,502 females. Public bus services are available within a 5-kilometer radius, and the nearest railway station is 17 kilometres away.

4.5.1 Cohort Study

Table 8: Cohort study of the village Kastadanga following Chandra, 2022 and Patra, 2022

Cohort				
Sl. No.	**Particulars/ Indicators**	**Years**		
		1990 - 2000	**2000 - 2010**	**2010 - 2020**
1	Crop Combination	Rice, jute, mustard, brinjal, parwal, chickpea and lentil.	Rice, jute, cauliflower, chilli, bitter gourd and flat beans.	Flat beans, boro rice (cultivation increased), jute cultivation decreased and has been replaced by vegetables.
2	Rice Yield (per bigha)	25 mon (1 mon = 40 kg)	21-22 mon	18-20 mon
3	Cost of Cultivation of Rice (Rs. per bigha)	4000	5500	10000
4	Soil Health	Very good	Medium	Poor
5	Water Availability	BhomraBil	BhomraBil	Shallow
6	Tree and crop biodiversity	Tree: Shishu, Aswatta, Banyan. Rice variety: Jaya, IR50, Ratna, Lal Swarna.	Tree: Eucalyptus, Mahogany. Rice variety: Ranjit, Pankaj.	Tree: Lombu, Kadam. Rice variety: Satabdi, Gosai.
7	Application of Organic Manure (cow dung)	More	Medium	Very less
8	Livestock Count (cows)	1500	600	30
9	No. of ponds	112	122	135
10	Rainfall (Rainy Days)	High (4-5 months)	Medium (3-4 months)	Less (2 months)

Cohort				
Sl. No.	**Particulars/ Indicators**	**Years**		
		1990 - 2000	**2000 - 2010**	**2010 - 2020**
11	Income/Net Return (Rs. per month)	3000	5000	8000
12	Expenditure (Rs. per month)	1000	4000	6500
13	Well-being (Score out of 10)	5	7	8

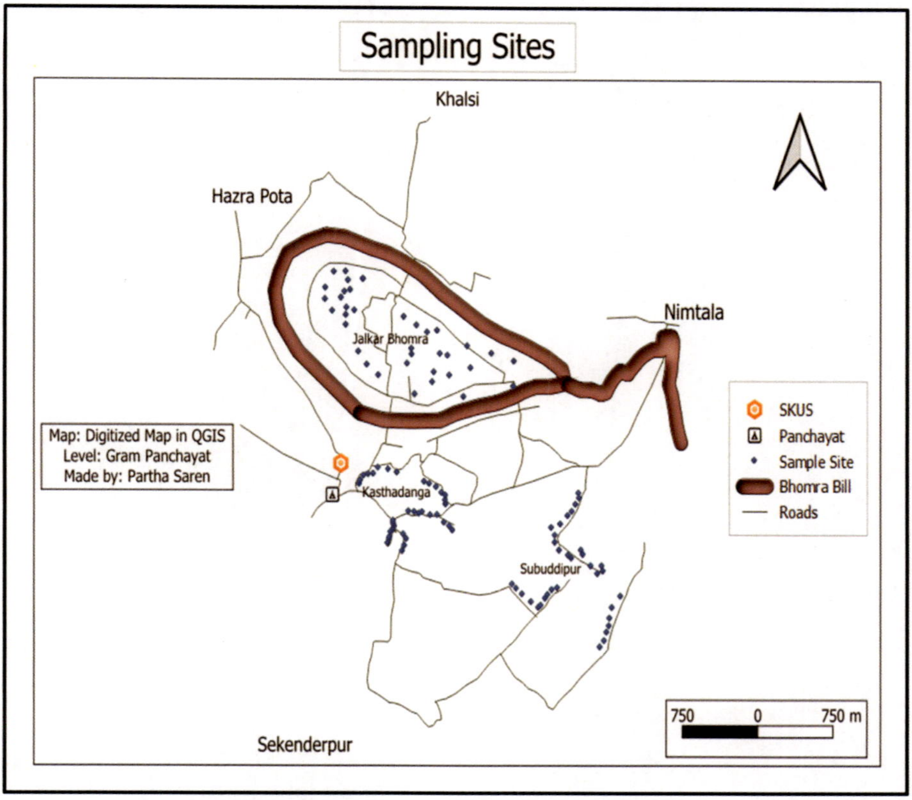

Map 5: Digitized Map of Sampling Sites

5

Materials and Method

The research methodology constitutes a comprehensive and well-structured plan of action for conducting the investigation and serves as a blueprint for the research process. Its purpose is to gain an understanding of the underlying concepts, methods, and techniques employed to define the study, gather relevant information, analyse data, and interpret the findings, thereby revealing concrete evidence and formulating theories.

In this chapter, an in-depth exploration of the research methodology is undertaken to comprehend the concepts, methods, and techniques employed for designing the study, gathering data, conducting data analysis, and interpreting the results. To facilitate clarity and ease of comprehension, the chapter has been divided into the following sub-headings:

1. Research Design
2. Locale of research
3. Pilot study
4. Sampling Design
5. Variables and their Empirical measurement
6. Methods of Data collection
7. Preparation of Interview Schedule
8. Item analysis and Expert's rating for devising scoring technique for variables
9. Pre-testing of Interview Schedule
10. Data Processing and Analysis for FCM
11. Statistical Tools used for Data Analysis
12. Software used for Data Analysis

5.1. Research Design

A research design provides a structured approach for researchers to answer their research questions in an unbiased, precise, and cost-effective way. It establishes a framework for testing relationships between variables and ensures that the research is going on right direction. There are various types of research designs available, and the current study uses the descriptive and diagnostic research design. Descriptive research design is used to collect accurate and relevant information about the current state of a phenomenon, and to draw valid and generalized conclusions wherever possible.

5.2. Selection of Research Locale

The current research was conducted in a specific agro-ecological zone within the state of West Bengal. The selected villages were located in the Haringhata block of the Nadia district, which falls under the new alluvial agro-ecological zone. The study took place between December, 2022 and May, 2023.

The area was selected because of several factors such as:

- The researcher had some understanding of the area, including its people, officials, and local dialects.
- The proximity of the area to the extension wing of Bidhan Chandra Krishi Viswavidyalaya provided opportunities to gather relevant data.
- The respondents in the area were highly cooperative and responsive, making it easier to collect information.

Additionally, the area was selected because it can be characterized by:

- Experienced and knowledgeable farmers who were willing to take risks and venture into new practices.
- The potential for establishing sustainable and profitable farming opportunities in the region.

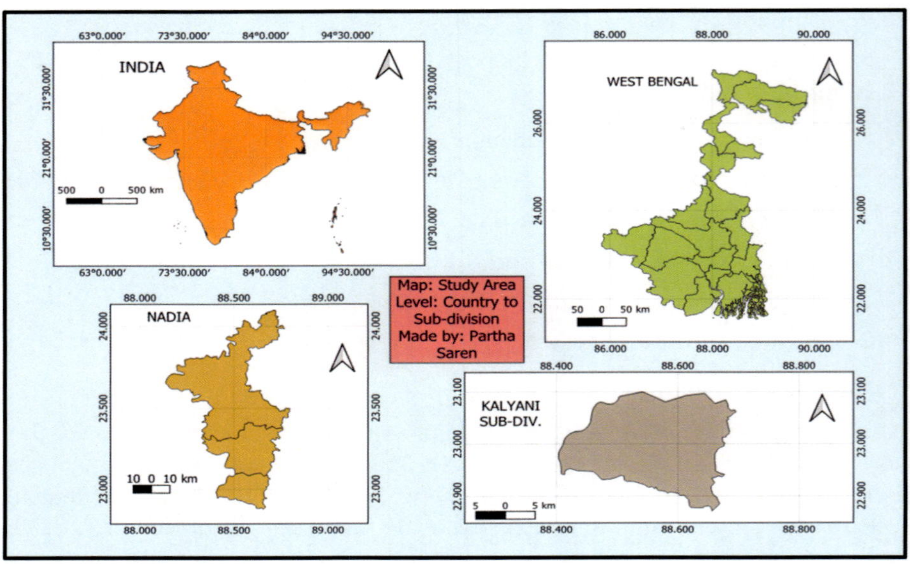

Map 6: Maps of Study Area

5.3. Pilot Study

A preliminary study was carried out on a small scale before undertaking the main study to gather initial information, which would serve as the foundation for planning the actual research work.

Pilot study had been carried out in the month of December, 2022 in the proposed research locale in order to find out whether the necessary data can be obtained or not. During this survey, informal discussions were held with farmers, local leaders, extension agencies, and state agricultural officials in the area. The purpose of these discussions was to gain insight into the socio-economic background of the farmers in the villages, their interest in sustainable agricultural practices, their experiences with sustainable farming, and their management strategies for dealing with the challenges.

The information collected from these discussions was used to prepare and refine the interview schedule, which was then finalized for use in the main study.

5.4. Sampling Design

Multi-stage random sampling procedure had been followed in the present study. The state West Bengal, the district Nadia and the block Haringhata were being purposively selected in the first, second and third stage respectively. In the fourth stage, three villages under the block namely, Subuddipur, Kastadanga and JalkarBhomra were being selected randomly. A total number of 90 respondents, 30 from each villages had been selected as representative sample for the study through random followed by snowball sampling method.

Table 9: Sampling Design

Stage	Locale	Approach	Justification
State (1)	West Bengal	Purposive	Because of the limitation of time and resources.
District (1)	Nadia	Purposive	Area with high fertile lands and high cropping intensity falls under new alluvial zone and chances of high natural resource degradation in the region.
Block (1)	Haringhata	Purposive	This block is one of the most agriculturally advanced block experiencing high cropping intensity and degradation of natural resources.
Village (3)	JalkarBhomra, Kastadanga, Subuddipur	Random	Out of the villages where farmers were growing more crops, this villages were selected randomly.
Farmers (90)	-do-	Random and Snowball	Three respondents, one from each village were randomly selected at first from a list of Contact Farmers. Afterwards, rest respondents were selected through snowballing method.

5.5. Variables and their Empirical Measurements

After reviewing various literatures related to the field of study and consultation with respected Chairman and Members of the Advisory Committee and other experts, a list of variables was prepared.

Table 10: Variables and their Empirical Measurements

Independent Variables		
Variable	**Notation**	**Empirical Measurement**
Age	x_1	Chronological age
Education	x_2	Scoring and Normalization
Functional education	x_3	No. of training and exposure to non-formal education
Size of family	x_4	Count
Cultivated Land Holding per capita	x_5	Bigha
No. of land fragment	x_6	Count
Avg. size of land fragment	x_7	Bigha
Distance between land fragment	x_8	Metre
Homestead land per capita	x_9	Katha
Marketed Surplus per capita	x_{10}	Tonne/year
On farm income per capita	x_{11}	Rupees/year
Family Expenditure per capita	x_{12}	Rupees/year
Savings per capita	x_{13}	Rupees/year
Land under irrigation per capita	x_{14}	Bigha
Cropping intensity	x_{15}	(GCA/NCA)x100
Livestock count	x_{16}	Count
Communication variable	x_{17}	Scoring and Normalization
LPG Consumption per person	x_{18}	Litre/year
Production of organic manure	x_{19}	Tonne/year
Application of organic manure	x_{20}	Tonne/year
Rice stubble height	x_{21}	Inches
Crop residue left in field	x_{22}	Kilogram/bigha/year
Soil conservation perception	x_{23}	Scoring and Normalization
Water conservation perception	x_{24}	Scoring and Normalization
Bio-diversity conservation perception	x_{25}	Scoring and Normalization
Application of climate resilient technology	x_{26}	Scoring and Normalization
Dependent Variable		
Perception on Climate Change	y_1	Index by means of scoring followed by calculation and Z standardization
Perceived Uncertainty	y_2	Index by means of scoring followed by calculation and Z standardization
Change Dynamics	y_3	Index by means of scoring followed by calculation and Z standardization

Soil conservation perception, Water conservation perception, Bio-diversity conservation perception, Application of climate resilient technology, Perception on Climate Change, Perceived Uncertainty and Change dynamics, all have been measured in terms of selected statements/items. These statements/items were selected after Item Analysis and Expert's Rating on 10 point scale based on the Summated Rating Scale (Likert's Scale).

5.6. Method of Data Collection

The current study requires a combination of quantitative and qualitative data. To gather this information, interviews will be conducted in a face-to-face setting, where the questions will be presented orally to the participants, and their responses will be recorded. This method allows for a direct interaction between the interviewer and the respondents, enabling a more comprehensive understanding of their perspectives.

To ensure consistency and enhance the suitability of the data for statistical analysis, a structured interview schedule will be employed. This schedule will consist of carefully designed and specific questions tailored to the requirements of the study. By using a structured approach, the data obtained through interviews can be easily compared and subjected to rigorous statistical analysis. Besides, interview offers a unique opportunity to delve deeper into the participants' experiences, thoughts, and insights, allowing for a more nuanced understanding of the research topic. Additionally, conducting interviews in person facilitates effective communication and enables the researcher to clarify any ambiguities or obtain additional information, contributing to the richness and depth of the collected data.Top of Form

The researcher personally conducted the data collection process by directly engaging with the farmers residing in the selected villages of Kastadanga-I gram panchayat during the month of February, March and April of 2023. A list contact farmers was obtained from the Kastadanga Gram Samabay Krishi UnnayanSamiti Limited from which three respondents, one from each village were randomly selected at first. Afterwards, rest respondents were selected through snowballing method.

5.7. Preparation of Interview Schedule

The primary tool used for data collection was an interview schedule. An interview schedule can be described as a formal document that serves as a systematic and standardized tool for conducting inquiries. Its primary function is to facilitate the collection of both quantitative and qualitative data in a cross-sectional manner.

To ensure a comprehensive and thorough data collection process, the interview schedule was carefully designed to include a combination of structured questions as well as open-ended inquiries. This approach aimed to gather pertinent information from the participants in a detailed manner.

5.8. Item Analysis and Expert's Rating for Devising Scoring Technique for Variables

Since the topic of the research is bit new and unique, the perception of farmers has been evaluated through some marker items/statements, generated by the researcher with the help and guidance of the experts in the relevant fields (Chairman and Members of the Advisory Committee of the Researcher) who provided valuable inputs. The steps involved in selection of marker item/statements, were as follows–

i. Selection of the experts and seeking their consents.

ii. To ensure a comprehensive assessment, a list of 60 randomly selected items was initially compiled.

iii. Screening and scrutinizing of the items based on applicability and relevancy.

iv. Finalization of marker item/statements to be rated on the Summated Rating Scale (Likert's Scale).

v. The experts were then asked to rate each item on a scale of 1 to 10, indicating the item's relevance to the specific aspects being considered. A score of 10 represented the highest level of relevance, while a score of 1 indicated the least relevance.

vi. Total score for each item was calculated.

vii. Around 48 percent of the items (29 items) were retained and rests were discarded.

viii. Out of which, 3 items were assigned to the each of the independent variable x_{23}, x_{24}, x_{25}; 5 items were assigned to the independent variable x_{26} and 3 items were assigned to the dependent variable y_1; 13 items were assigned to the dependent variable y_2, and 4 items were assigned to the dependent variable y_3.

ix. The final marker items/ statements had thus been finally selected and applied for data collection.

Likert's Scale: Likert's Scale of Summated Rating, also known as the Likert scale, is a commonly used psychometric tool for measuring attitudes, opinions, and perceptions of individuals. It was developed by Rensis Likert, an American social scientist, in the 1930s.

The Likert scale consists of a series of statements or items that express various degrees of agreement or disagreement with a particular statement. Typically, respondents are asked to indicate their level of agreement or disagreement by selecting a response option on a numerical scale. The scale usually ranges from 1 to 5 or 1 to 7, with 1 representing "strongly disagree" or "strongly negative" and the highest number representing "strongly agree" or "strongly positive." The scale may also include a midpoint option for those who are neutral or undecided.

Responses from multiple individuals are then aggregated and analysed to provide insights into the collective attitudes or opinions of a group. The scores obtained from the Likert scale can be used to compare different groups, track changes over time, or evaluate the effectiveness of interventions or treatments. The Likert scale is widely used in various fields, including social sciences, psychology, market research, and customer satisfaction surveys. It provides a structured and quantifiable approach to measure subjective constructs, allowing researchers to gain valuable insights into people's opinions and attitudes.

5.9. Pre-testing of Interview Schedule

Pretesting is a crucial process in research studies that involves advanced testing of the study design after the schedule has been prepared. Its main objective is to identify any discrepancies or issues that have arisen and address them through necessary modifications to the schedule.

In the pretesting phase, the interview schedule underwent appropriate changes and modifications based on the feedback received. The structured interview schedule was pre tested in a non-sampled area which includes different villages under Kastadanga-II gram panchayat in the month of January, 2023. Modifications, deletions, and additions were made to the interview schedule based on the feedback gathered during the pretesting phase. Finally, the interview schedule was deemed finalized and ready for implementation in the main study.

5.10. Data Processing and Analysis for FCM

'Mental Modeler' software was used to formulate relationships among FCM components which acts as an internal construct that give an interpretation of external components to make decision of the individual (Gray et al., 2013). With contributions from expert consultations, a Likert scale was selected to assess psychological behaviour of farmers with respect to decision-making (Burton, 2004). The scale was prepared in consultation with the experts of therelevant fields (Chairman and Members of the Advisory Committee of the Researcher). Farmers were asked a number of statements on the causal effect or relationship between the components of their farming system. Farmers' opinions on relationships of factors (if it is strong or weak) for their farming system was noted, using a Likert scale to measure the level of effectiveness as well as farmers' responses (Guillem et al., 2012).

The process of developing the FCMs was conducted through the collection of concepts from the respondents, developing a hypothetical concept map by drawing interconnections among relevant conceptual components using information from farmers. Components of the farming systems were characterized by different weights to describe the causal relationship between two concepts. Values of weights were in the range of [-1, 1], which was converted from the Likert scales. Influences of one concept on another concept were represented by the weight of the relationship. A positive relationship between two concepts indicates that an increase or decrease of the primary concept would result in the same directional influence on the secondary concept. Therefore, a negative relationship indicates an inverse relationship between two connected concepts. On the contrary, a neutral value of the relationship indicates that there is no perceived relationship between concepts (Stylios and Groumpos, 1999). The same interview schedule was used for data collection as in used to collect data for dependent and independent variables however used in different manner. The intention was to know their perception regarding the perception on climate change, level of uncertainty and trend of agricultural change.

Collected data were entered into a MS Excel sheet. To identify the value of relationships between concepts, the scores were converted to number value ranging from [-1 to 1]. As the range of weight of the value is limited within 'Mental Modeler' software, the scores were converted to the range from 1 to -1 through Minimum-Maximum (Min-Max) normalization.

Short description of components of 'Mental Modeler' software is stated below-

Component/Concept: Variables in the model.

Driver component: A variable that has influence on another variable.

Receiver component: A variable that is influenced by other variables.

Ordinary component: A variable having both driving and receiving functions.

5.11. Statistical Tools Used for Data Analysis

Table 11: Statistical Tools used for Data Analysis

Sl.No.	Approaches	Statistical Tool
1.	Descriptive visualization	Mean, Minimum, Maximum, Range, Standard Deviation, Coefficient of Variation
2.	Influence Analysis	Coefficient of Correlation
3.	Functionality Analysis	Multiple Regression and Stepwise Regression
4.	Direction Analysis	Path Analysis
5.	Dimension Reduction	Factor Analysis (Principal Component Analysis/PCA)
6.	Conjoint	Canonical Covariate Analysis (CCA)
7.	Multi-layer perceptron	Artificial Neural Network (ANN)

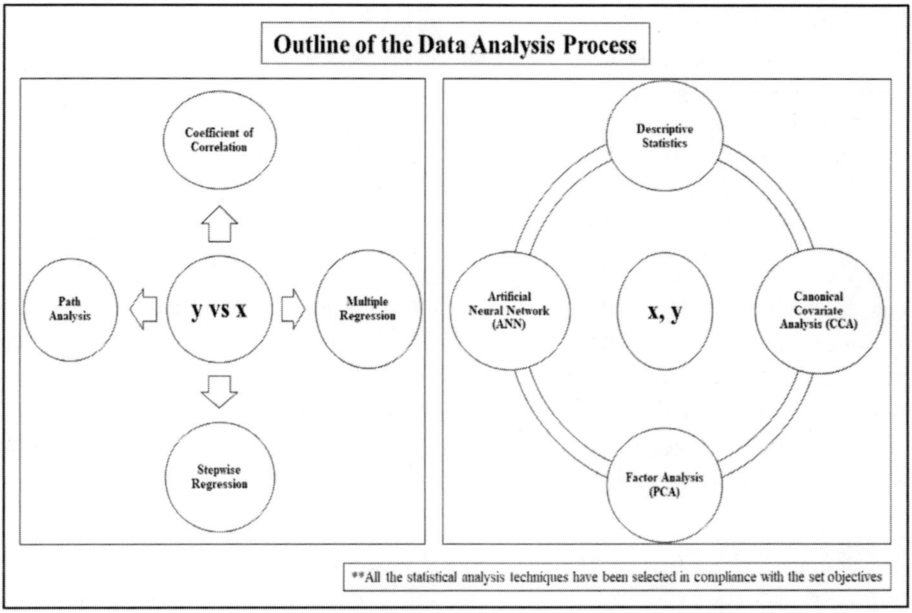

Fig. 27:

The statistical tools and techniques used in the present study for the analysis of the raw data are discussed in details as follows -

5.11.1 Mean

It is the arithmetic average and is the result obtained when the sum of the values of the individuals in the data is divided by the number of individuals in the data (Panse and Sukhatme, 1967). The mean is the most basic and comparatively consistent measure of central tendency. The mean reflects and is affected by every score in the distribution. For social action purpose, a mean may not provide a realistic picture of the situation.

The Mean of ungrouped data is computed by using the formula, $\overline{X} = \dfrac{\Sigma x}{N}$

Where, $\overline{X}$ = mean of the scores; Σx = sum of individual scores and N = number of observations.

When the data are expressed in a frequency distribution (grouped), the mean is calculated using the following formula,

$$\overline{X} = \frac{\Sigma f \times x}{N}$$

Where, $\overline{X}$ = mean of the scores; f = frequency of the class; x = class value or mid value of the class and N = number of observations.

5.11.2 Range

The Range of a distribution is the difference between the smallest and largest of observations and gives us some idea of the amount of variability present (Panse and Sukatme, 1967). The range is the most general measure of spread or scatter and may be computed to make a rough comparison of two or more groups for variability. The range takes into account the extremes of the series of scores only and no weightage is given for other observations.

5.11.3 Standard Deviation

The square root of the arithmetic mean of the all squared deviations, where the deviations are calculated from the arithmetic mean of the distribution is called the standard deviation. It is generally denoted by the symbol σ (sigma). It is less affected by sampling error and is a more stable measure of dispersion.

The standard deviation of the data grouped in the form of a frequency distribution is computed by the formula,

$$\sigma = \sqrt{\frac{\Sigma f.\,d^2}{N}}$$

Where, f = frequency of the class; d = deviation of the mid-value of the class from the population; N = total number of observations.

5.11.4 Coefficient of Variation

A measure of variation which is independent of the unit of measurement is provided by the Coefficient of variation. Being unit free, this is useful for comparison of variability between different populations. The coefficient of variation is standard deviation expressed as percentage of the mean. It is measured by the formula,

$$C.V. = \frac{Standard\ Deviation}{Mean} \times 100$$

5.11.5 Coefficient of Correlation

Correlation refers to the phenomenon of occurrence of increase or decrease in one variable when there is a corresponding increase or decrease in another variable. These variables are said to be correlated. The correlation coefficient (r) is a metric that quantifies the association between two variables that are measured at the interval or ratio level and exhibit a linear relationship.

Pearson product-moment r is computed by the formula,

$$r_{xy} = \frac{N\Sigma xy - (\Sigma x)(\Sigma y)}{\sqrt{\left[N\Sigma x^2 - (\Sigma x)^2\right]\left[N\Sigma y - (\Sigma y)^2\right]}} \frac{N\Sigma xy - (\Sigma x)(\Sigma y)}{\sqrt{\left[N\Sigma x^2 - (\Sigma x)^2\right]\left[N\Sigma y - (\Sigma y)^2\right]}}$$

Where, x and y = original scores in variables X and Y; N = number of paired scores; Σxy = each x multiplied by its corresponding y, then summed; Σx = sum of x scores; Σx^2 = each x squared, then summed; $(\Sigma x)^2$ = sum of x scores, squared; Σy = sum of y scores; Σy^2 = each y squared, then summed; $(\Sigma y)^2$ = sum of y scores, squared.

5.11.6 Multiple Regression

The underlying relation between y and x in a bivariate population can be expressed in the form of a mathematical equation known as Regression equation and is said to represent the regression of the variate y on the variate x (Panse and Sukhatme, 1967). If y is the dependent variable and x is the independent variable, then the linear regression equation can be written as $\boldsymbol{y = a + bx}$

Multiple regression is a statistical technique used to examine the relationship between a dependent variable and two or more independent variables. It allows researchers to assess the impact of multiple predictors simultaneously and understand how they collectively influence the dependent variable. Multiple regression provides insights into the strength, direction, and statistical significance of the relationships. The formula for multiple regression is represented as:

$$y = \beta_o + \beta_1 x_1 + \beta_2 x_2 + ... + \beta_1 x_1 + \varepsilon$$

Where, **y** is the dependent variable; $\mathbf{x_1, x_2, ..., x_1}$ are the independent variables; βo, β_1, β_2, ..., β are the regression coefficients (also known as beta weights or slopes) representing the contribution of each independent variable to the dependent variable; ε is the error term or residual, representing the unexplained variance in the dependent variable.

The goal of multiple regression is to estimate the regression coefficients (βo, β_1, β_2, ..., β) that minimize the difference between the observed values of the dependent variable (y) and the predicted values based on the independent variables (x_1, x_2, ..., x). This is typically done using the method of least squares, which minimizes the sum of squared residuals. The estimated regression coefficients are obtained through calculations involving the independent and dependent variables, as well as their means and variances. The formula for calculating the regression coefficient (β) is:

$$\boldsymbol{\beta = (\Sigma(x -) (y -)) / (\Sigma(x -)^2)}$$

Where, x is an independent variable; y is the dependent variable; is the mean of x; is the mean of y.

The multiple regression analysis also assesses the overall fit of the model and the statistical significance of the regression coefficients. Some key statistical measures include:

R-squared (R^2): This measure indicates the proportion of variance in the dependent variable that can be explained by the independent variables. It ranges from 0 to 1, with higher values indicating a better fit of the model.

Adjusted R-squared: This adjusted version of R-squared takes into account the number of predictors and the sample size, penalizing the inclusion of unnecessary variables in the model.

F-statistic: This statistic assesses the overall significance of the regression model. It tests the null hypothesis that all regression coefficients are zero simultaneously. A significant F-statistic suggests that at least one independent variable has a significant impact on the dependent variable.

p-values: Each regression coefficient has an associated p-value, which indicates the statistical significance of the individual predictor. A p-value below a predetermined significance level (e.g., 0.05) suggests a significant relationship between the independent variable and the dependent variable.

In addition to these measures, researchers may also examine the standard errors of the regression coefficients, confidence intervals, and collinearity diagnostics to assess the stability and reliability of the model. Multiple regression analysis provides valuable insights into the relationships between variables, allowing researchers to make predictions and understand the factors that influence the dependent variable. It is widely used in various fields, including social sciences, economics, business, and healthcare research.

5.11.7 Stepwise Regression

Stepwise regression is a variation of multiple regression which provides a means of choosing independent variables that yield the best prediction possible with the fewest independent variables. It permits the user to solve a sequence of one or more multiple linear regression problems by stepwise application of the least square method. At each step in the analysis, a variable is added or removed which results in the greatest production in the error sum of squares (Burroughs Corporation, 1975).

According to Draper and Smith (1981), the method of stepwise multiple regression analysis is to insert variables in turn until the regression equation is satisfactory. The order of insertion is determined by using the partial correlation coefficient as a measure of the importance of variables not yet in the equation. Mainly it is used to select a subset of independent variables that best predict the dependent variable. It systematically examines a set of predictors and determines their contribution to the regression model based on statistical criteria. Stepwise regression aims to improve the predictive accuracy of the model while maintaining parsimony. There are two main approaches to stepwise regression: forward selection and backward elimination:

Forward Selection: In forward selection, the analysis starts with an empty model and iteratively adds predictors one by one. At each step, the predictor that contributes the most to the model is selected based on a statistical criterion, such as the F-statistic or t-statistic. The addition of each predictor is evaluated based on a predetermined criterion, such as a significance level (e.g., p-value < 0.05) or an increase in R-squared value. If the criterion is met, the predictor is retained in the model; otherwise, it is excluded.

Backward Elimination: In backward elimination, the analysis starts with a model that includes all predictors. At each step, the predictor that contributes the least to the model is removed based on a statistical criterion. The elimination of each predictor is evaluated based on a predetermined criterion, such as a significance level (e.g., p-value > 0.05) or a decrease in R-squared value. If the criterion is met, the predictor is removed from the model; otherwise, it is retained.

The stepwise regression procedure continues until no further predictors meet the inclusion or elimination criteria, resulting in the final model. There are several statistical criteria commonly used to determine whether a predictor should be included or excluded in stepwise regression:

F-Statistic: This criterion is often used in forward selection. At each step, the predictor with the highest F-statistic (indicating the largest improvement in model fit) is selected for inclusion in the model.

t-Statistic: This criterion is often used in backward elimination. At each step, the predictor with the lowest t-statistic (indicating the least contribution to the model) is eliminated.

p-Value: This criterion assesses the statistical significance of each predictor. A predetermined significance level (e.g., p-value < 0.05) is used to determine whether a predictor should be included or excluded.

Change in R-Squared: This criterion evaluates the increase or decrease in the R-squared value of the model. A significant increase in R-squared suggests that adding a predictor improves the model's predictive accuracy.

While stepwise regression can be useful for variable selection, it is important to interpret the results with caution and consider the limitations of the approach. Stepwise regression may lead to overfitting or the inclusion of spurious predictors. It is advisable to validate the final model using independent datasets or perform cross-validation techniques to ensure its generalizability. Stepwise regression can aid in model parsimony and improve predictive accuracy, but it should be used judiciously and in combination with other validation techniques.

5.11.8 Path Analysis

Path Analysis is done to get a clear picture of the direct and indirect effects of the independent variables on the dependent variable. Variables through which substantial indirect effects are channelled, also found out. Singh and Chaudhary (1977), defined path coefficient as the ratio of the standard deviation of the effect due to a given cause to the total standard deviation of the effect i.e. if y is the effect and x_1 is the cause, the path coefficient for the path from cause x_1 to the effect y is -

$\sigma x_1/\sigma_y$

The primary goal of path analysis is to assess the causal pathways through which variables influence each other. It helps researchers to test theoretical models and hypotheses about how different factors contribute to an outcome of interest. Path analysis enables the investigation of both direct and indirect effects of independent variables on a dependent variable, taking into account the mediating variables that may lie in the causal chain.

To conduct a path analysis, researchers typically begin by developing a theoretical model that outlines the relationships between variables based on prior knowledge or existing theories. The model consists of a set of variables and directional arrows that represent the expected relationships between them. The variables can be observed or latent (unobserved) constructs. Once the model is defined, researchers collect data on the variables of interest. These variables can be measured using various methods such as surveys, questionnaires, observations, or existing data sources. The data are then analysed using specialized software that allows for path analysis, such as OPSTAT, structural equation modelling (SEM) software etc.

The analysis begins by estimating the relationships between variables using regression techniques. The researcher examines the strength and significance of the direct effects of independent variables on the dependent variable. Additionally, path analysis allows for the assessment of indirect effects, which are the effects that occur through one or more intermediate variables in the causal chain. Indirect effects are particularly useful in path analysis because they help to uncover the underlying mechanisms by which variables influence each other. Mediation analysis, a subfield of path analysis, focuses specifically on identifying and quantifying indirect effects. It allows researchers to determine whether the relationship between two variables is mediated by a third variable, providing insight into the causal processes at work. Path analysis also enables researchers to evaluate the overall fit of the model to the data. Goodness-of-fit indices, such as the chi-square test, comparative fit index (CFI), and root mean square error of approximation (RMSEA), are used to assess how well the model matches the observed data. If the model does not fit well, modifications can be made to improve its fit, such as adding or removing paths between variables.

By using path analysis, researchers can gain a deeper understanding of the complex relationships among variables and identify the most influential factors contributing to an outcome of interest. It allows for the examination of both direct and indirect effects, providing a comprehensive picture of the underlying processes. Path analysis is widely used in social sciences, psychology, education, and other fields where understanding causal relationships is essential for theory development and practical applications.

5.11.9 Factor Analysis

Factor analysis is a very effective and popular multivariate analysis method that is typically used in the social and behavioural sciences. It is a set of analytical techniques intended to condense data by grouping variables based on their inter-correlations or shared variance. The underlying assumption is that variables showing correlation likely measure a similar factor, trait, or construct. For smaller datasets or variables with clear content, researchers can visually identify groupings. However, this becomes challenging for larger datasets or ambiguous items. In such cases, factor analysis is employed to reduce the information in a model by reducing the dimensions of observations. Factor analysis serves multiple purposes. It can simplify data by reducing the number of variables in predictive regression models. In such cases, factors are often rotated after extraction. Various rotation methods exist, including those ensuring orthogonal factors, where the correlation coefficient between two factors is zero, thus addressing multi-collinearity issues in regression analysis. Factor analysis is also useful in theory testing to validate scale construction and operationalization. In this scenario, the scale is predetermined, and a subset of the scale is expected to represent an independent dimension within it. This form of factor analysis is commonly utilized in structural equation modelling (SEM) and is known as Confirmatory Factor Analysis. Additionally, factor analysis can be employed in constructing indices. While the simplest method is summing up the items in an index, there are situations where certain variables hold greater explanatory power or where similar questions are highly correlated, allowing for the removal of redundant questions to shorten questionnaires. Factor analysis can assist in determining the weight each variable should carry in the index.

Factor analysis is a data reduction technique that groups items together, allowing for a better understanding of the structure or composition of each group and providing a more comprehensive explanation of the data. It is important to note that factor analysis does not aim to predict or explain relationships between different sets of variables or determine group differences. Instead, it focuses solely on one set of variables and seeks to explain the underlying structure or composition of the data. There are two types of factor analysis: exploratory factor analysis (EFA) and confirmatory factor analysis (CFA). EFA is used to explore or derive the underlying factor structure of a data matrix, often without relying on pre-existing theories. EFA includes two commonly used types: principal components analysis (PCA) and principal factor analysis (PFA), with PFA preferred by most journals currently. On the other hand, CFA is employed to test existing theories. The researcher specifies the factors and the variables/items constituting each factor, then assesses how well these parameters fit the data. CFA is typically conducted using structural equation modelling (SEM). The subsequent discussion focuses on factor analysis related to EFA.

The primary goal of factor analysis is to explain as much variance as possible in a data set by employing the fewest factors and minimizing the number of items or variables within each factor. The balance between explained variance and simplicity is crucial. It is essential to ensure that the remaining variance in the solution primarily represents error variance. In EFA, the first factor derived is a linear combination of all variables/items that maximizes the explained (or extracted) variance. No other linear combination can achieve a higher variance. The proportion of variance extracted is known as the eigenvalue. The process continues, extracting subsequent factors, each orthogonal (uncorrelated) to the previous one, until all variance is extracted, resulting in the same number of extractions as the original variables/items.

Interpretation of factor analysis involves three steps:

1. Determining the number of factors to retain. Various decision rules can be utilized, but typically factors with eigenvalues greater than 1.0 are retained. The rationale is that a variable has a variance equal to 1.0, and eigenvalues should explain more variance than at least one variable. In a satisfactory EFA, the total variance explained by the retained eigenvalues should exceed 70%.

2. Identifying which variables/items "load" on each factor. This is accomplished by examining the coefficient of each variable/item, representing its correlation with the factor. Each variable/item will have a different coefficient for each factor since each factor is a linear combination of all variables/items. The higher the coefficient, the greater the contribution of the variable/item to the factor, and the square of the coefficient represents the amount of variance explained by the variable/item. Ideally, each variable/item should have a substantial loading on only one factor and smaller loadings on the remaining factors.

Before determining the composition of each factor, the factor dimensions can be geometrically rotated, positioning the new set of axes closer to their respective factor variables/items. These axes are similar to those in a Cartesian coordinate system. For instance, in a two-factor model, each variable/item will have a factor loading value for Factor I and Factor II. Most variables/items will exhibit a relatively high loading value (>0.5) on one factor and a smaller loading on the other (<0.5). Each factor represents an axis in space, and the factor loading of a variable/item is its coordinate value on that axis. By rotating the axes, the value along the Factor I dimension can be increased while decreasing the Factor II dimension value, as long as the distance of the variable/item to the origin remains unchanged. This strengthens the association between the variable/item and Factor I while reducing its association with Factor II, simplifying the interpretation of the factor structure. Orthogonal rotation is commonly assumed when rotating the axes, where the axes remain at a 90-degree angle to one another. In a two-factor model, if Factor I is rotated 35 degrees clockwise, Factor II must also be rotated 35 degrees clockwise. However, if the factors are somewhat correlated, the rotation is based on the correlation between them. The cosine of the angle between the axes equals the correlation. This rule applies to orthogonal rotation because the angles are 90 degrees, and the cosine of 90 degrees is 0.

3. Organizing the output into a table with columns representing the factors and rows representing the variables/items. Loadings greater than 0.50 for each factor are highlighted or underscored. Once all contributing variables across all factors have been identified, the content of each factor is determined by analysing the general theme or content of the highlighted variables/items, and an appropriate name is assigned to each factor.

5.11.10 Canonical Covariate Analysis

Canonical-Correlation Analysis (CCA) also called Canonical Variates Analysis or Covariate Analysis, is a way of inferring information from cross-covariance matrices. A canonical correlation is the correlation of two canonical variables, one representing a set of independent variables, the other a set of dependent variables. Each set may be considered a latent variable based on measured indicator variables in its set. The method was first introduced by Harold Hotelling in 1936 although in the context of 'angles between flats' the mathematical concept was published by Jordan in 1875.

Canonical correlation is a member of multiple general linear hypothesis (MLGH) family and shares many of assumptions of multiple regression such as linearity of relationship, homoscedasticity (same level of relationship for the full range of data), interval or near interval data, untruncated variables, proper specification of model, lack of high multi-collinearity and multivariate normality for purpose of hypothesis testing. If we have two vectors $X = (X_1 .... X_n)$ and $Y = (Y_1 ..... Y_m)$ of random variables and there are correlations among the variables, then canonical-correlation analysis will find linear combinations of X **and** Y **which have maximum correlation with each other.**

T. R. Knapp notes that "virtually all of the commonly encountered parametric tests of significance can be treated as special cases of canonical-correlation analysis which is the general procedure for investigating the relationships between two sets of variables."

The canonical correlation is optimized such that the linear correlation between the two latent variables is maximized. Canonical correlation is used for many to many relationships. There may be more than one such linear correlation relating the two set of variables, with each such correlation representing a different dimension by which the independent set of variables is related to the dependent set.

The purpose of canonical correlation is to explain the relation of the two sets of variables not to model the individual variables. In addition to asking how strong the relationship is between two latent variables, canonical correlation is useful in determining how many dimensions are needed to account for that relationship. Canonical correlation finds the linear combination of variables that produces the largest correlation with second set of variables.

5.11.11 Artificial Neural Network

Artificial neural networks (ANNs), commonly referred to as neural networks (NNs), are computational systems based on machine learning algorithm that draw inspiration from the biological neural networks found in animal brains.

A neural network comprises an input layer consisting of neurons (or units), one or two hidden layers of neurons (or units), and a final layer of output neurons. Each connection between neurons is characterized by a weight, a numerical value.

The output, h_i, of neuron i in the hidden layer is determined by the equation:

$$A_i^{(t+1)} \frac{x-\mu}{\sigma}$$

Here, σ represents the activation (or transfer) function, N denotes the number of input neurons, V_{ij} represents the weights and X_j represents the inputs to the input neurons and represents the threshold terms of the hidden neurons.

The activation function serves two purposes: introducing nonlinearity into the neural network and ensuring that the neuron values do not diverge, preventing the network from becoming paralyzed. It has been demonstrated that a neural network constructed as described above can approximate any computable function with arbitrary precision. The inputs provided to the input neurons are independent variables, while the outputs returned by the output neurons are dependent variables in relation to the function being approximated by the neural network. When appropriately encoded, the inputs to and outputs from a neural network can be binary (such as yes or no) or even symbolic (e.g., green, red...), which enhances the versatility of neural networks.

Training is another crucial aspect of neural network applications. Similar to human learning through examples, a neural network is trained by exposing it to a set of input data known as the *training set*. The desired outputs for the training data are known, and the goal of training is to minimize the error function by adjusting the weights between connected neurons. The *error function* typically consists of the sum of squared differences between the neural network outputs and the desired outputs.

When experimenting with different neural network architectures, an independent data set called a *validation set* can be utilized to evaluate the performance of trained neural networks. The best-performing architecture is selected as the preferred choice. Following the validation process, an additional independent data set known as the *test set* is employed to assess the performance level of the neural network, providing an indication of its reliability. It should be noted that a neural network can only learn what is present in the training set. Therefore, the size of the training set needs to be sufficiently large for the neural network to capture the embedded features and trends. Conversely, if the training set contains excessive unimportant details, the neural network might allocate resources (weights) to fitting the noise. Consequently, a judicious selection and representation of data are critical for the successful implementation of neural networks.

5.11.12 Mental Modelling: Fuzzy-logic Cognitive Mapping

The semantics behind a standard FCM can be defined by a 4-tuple (C, W, A, *f*), where

C = $\{C_1, C_2 \ldots\ldots C_M\}$ is the family of *M* concepts modeled after fuzzy sets;

W: C × C → [-1, 1] is the matrix containing the weight $w_{ij} \in [-1, 1]$ assigned to each pair of concepts (C_i, C_j). The value of w_{ij} determines the sign and intensity(magnitude) of the edge connecting the cause concept C_i with the effect conceptC_j.

The function A: C → $A_i^{(t)}$ computes the activation degree $A_i \in R$ of each concept C_i at the discrete time step $t = \{1, 2\ldots\ldots T\}$.

Finally, the transfer function *f*: R → *I* aggregates the impact of multiple causal events over the target concept and clamps the result to the predefined activation interval.

The interpretation of the causal weight w_{ij} between two concepts C_i and C_j is as follows:

If $w_{ij}>0$, then an increment (decrement) in the concept C_i will produce an increment (decrement)on concept C_j *with* intensity $|w_{ij}|$.

If $w_{ij}<0$, then an increment (decrement) in the concept C_i will produce a decrement (increment) on conceptC_j with intensity $|w_{ij}|$.

If $w_{ij} = 0$ (or very close to 0), this denotes the absence of a causal relationship from C_i upon C_j, so there is no causal edge in the graph.

The adjacency matrix

The adjacency matrix W is defined by W = (w_{ij}), where w_{ij} is the weight of directed edge $C_{i,}$ C_j. Clearly, W is a (n × n) matrix, where the diagonalentries are zero, since the weight of the edge $C_{i,}$ C_j does not exist.

The Activation Function

Equation (1) showsKosko's activation rule for FCMs, with $A^{(0)}$ being the initial configuration (activation vector), w_{ji} the value of the causal relation connecting concept C_j to concept C_i whereas denotes the activation value of concept C_i at the *t*-th time step. This update rule is iteratively repeated until a stop condition is met. A FCM will produce a state vector at each discrete time step that comprises the activation degree of all concepts.

$$A_i^{(t+1)} = f\left(\sum_{\substack{j=1 \\ i \neq j}}^{M} wjiA_j^{(t)}\right) \tag{1}$$

The above equation describes an inference (updating) rule widely used in manyFCM-based applications, but it is not the only one. Stylios and Groumpos proposed a modified inference rule where concepts take into account their own past activation value besides the corresponding weights and activation values coming from other concepts. This reasoning rule is preferred when updating concepts that are not influenced by other concepts. The reader can notice that this implicitly removes the i≠jconstraint in Kosko's equation.

$$A_i^{(t+1)} = f\left(\sum_{\substack{j=1 \\ i \neq j}}^{M} wjiA_j^{(t)} + A_i^{(t)} \right) \quad (2)$$

Selecting the proper update rule depends on the problem at hand and regularly requires a strong understanding of the physical/simulated system under investigation. As a further valuable remark, Papakostas and Koulouriot is observed that removing the i≠jrestriction in Equations (1) and (2) does not necessarilyimprove the overall prediction rates of FCM-based classifiers.

One of the most relevant characteristics of FCMs is the interpretability of their topology. In fact, FCMs can be defined as interpretable recurrent neural networks that include fuzzy logic elements during the knowledge engineering phase. More explicitly, an FCM exploits an activation vector by using a rule similar to the standard McCulloch-Pitts model where concepts can be thought of as neural processing entities. This implies that the activation degree of each map neuron is given by the value of the transformed weighted sum this processing unit receives from the connected neurons in the causal network.

The transfer function

The function f: R → I in Equations (1) and (2) denotes a monotonically, non-decreasing function used to clamp the activation value of each concept to the desired interval I, where $I = [0, 1]$ or $I = [-1, 1]$ depending on the problem domain.

According to the cardinality of the state space, existing transfer functions may be categorized into either *discrete* or *continuous*. The most widely used transfer functions are the bivalent, the trivalent, the hyperbolic tangent and the sigmoid function.

In the present research work, the hyperbolic tangent transfer function has been used.

The hyperbolic function is a continuous transfer function that produces infinite states that are freely distributed within the $[-1, 1]^M$hypercube. Besides the equilibrium points and the cyclic states, continuous functions might additionally yield chaotic outputs. However, they can be used for modeling both qualitative and quantitative scenarios.

It can be mathematically expressed by the following equation

$$f3(x) = \frac{e^{2x} - 1}{e^{2x} + 1}$$

5.12 Software Used for Data Analysis

Table 12: Software used for data analysis

Sl.No.	Analysis/Approach	Software
1.	Descriptive and Inferential Statistics	• Statistical Packages for Social Sciences (SPSS) • MS-Excel • OPSTAT
2.	Mental modelling – Fuzzy Logic Cognitive Mapping	Mental Modeler
3.	Study area and sampling site mapping	QGIS

5.12.1 SPSS

Statistical Packages for Social Sciences or commonly known as the SPSS is a statistical software suite developed by IBM for data management, advanced analytics, multivariate analysis, business intelligence, and criminal investigation. Long produced by SPSS Inc., it was acquired by IBM in 2009. Current versions (post 2015) have the brand name: IBM SPSS Statistics.

SPSS offers advanced statistical analysis, a vast library of machine-learning algorithms, text analysis, open-source extensibility, integration with big data and seamless deployment into applications. It is one of the widely used statistical packages in the field of social science and human behaviour research. Descriptive statistics, various dimension reduction and classification techniques, parametric as well as non-parametric tests, regression and correlation analysis tests etc. can be performed with ease by dint of this user-friendly software.

IBM SPSS Statistics version 25 was used in the data analysis of the present study.

5.12.2 MS-Excel

Microsoft Excel is a spreadsheet developed by Microsoft. It features calculation or computation capabilities, graphing tools, pivot tables and a macro programming language called Visual Basic for Applications (VBA). Excel forms the part of Microsoft 365 suite of software.

There are plenty of statistical functions and data analysis tools available in MS Excel. Descriptive statistics, fitting of distributions and some important parametric tests (t test, F test etc.) can be done with statistical functions. However, by using 'Analysis Tool Pack' of MS Excel, ANOVA for single as well as multiple factors, regression, data summary can be computed directly for the given set of data.

MS-Excel version 2016 (Professional Plus) was used in the data analysis of the present study.

5.12.3 Opstat

OPSTAT is a free web based software for statistical analysis developed at Chaudhary Charan Singh Haryana Agricultural University, Hisar, Haryana, India.

This tool is particularly developed for agriculture data analysis by O. P. Sheoran, Computer Programmer at CCS HAU, Hisar, India. One factor analysis, two factor analysis, three factor analysis, correlation and regression analysis, generation means analysis, diallel analysis, path analysis, line × tester analysis, partial diallel analysis and stability analysis can be easily done with OPSTAT.

5.12.4 Mental Modeler

Mental Modeler is a decision-support software intended to help individuals and communities by means of Scenario Analysis so to understand the impacts associated with environmental change and develop mitigation strategies to reduce unwanted outcomes by capturing, communicating and representing knowledge.

Through a multi-step process based in Fuzzy-logic Cognitive Mapping (FCM), Mental Modeler allows groups of stakeholders to come together and easily develop semi-quantitative models of environmental issues which: (i) define the important components relevant to a community, (ii) define the strength of relationships between these components and (iii) run "what if" scenarios on these models to determine how the system might react under a range of possible conditions.

5.12.5 QGIS

QGIS, an open-source desktop geographic information system (GIS) application, is freely available across different platforms. It offers a range of functionalities, including the ability to view, edit, print, and analyse geospatial data. The development of the software, initially known as Quantum GIS, commenced in early 2002 under the guidance of Gary Sherman. In 2007, it was adopted as an incubator project by the Open Source Geospatial Foundation. The first official release, version 1.0, was made available in January 2009. To prevent confusion caused by the simultaneous use of the names "Quantum GIS" and "QGIS", the official name was changed to QGIS when version 2.0 was released in 2013.

QGIS empowers users to analyse and edit spatial information and provides the capability to create and export graphical maps. It supports various types of data layers, including raster, vector, and mesh layers. Vector data can be stored as points, lines, or polygons, while multiple formats of raster images are compatible, with the ability to georeference images. Additionally, QGIS supports industry-standard formats such as shape files, personal geodatabases, dxf, MapInfo, and PostGIS. It also facilitates the utilization of data from external sources through support for web services like Web Map Service and Web Feature Service. Moreover, QGIS is compatible with SAGA GIS and Kosmo for further flexibility in its usage.

QGIS version 3.28 was used in the present study.

6

Empirical Study

This chapter focuses on presenting empirical findings from data analysis to provide sufficient interpretation so to draw reliable conclusions. To enhance and complement the textual contents of the study, tables and models are frequently utilized. Additionally, the discussion serves to contextualize the study's findings and establish connections with the research work.

6.1 Descriptive Statistics of Variables With Respect to Range, Minimum, Maximum, Mean, Standard Deviation of Values, Variance, and Coefficient of Variation

Table 13: Descriptive Statistics of Variables

Variable	Range	Minimum	Maximum	Mean	Std. Deviation	Variance	CV (%)
Age (x_1)	44.00	26.00	70.00	42.98	10.77	115.96	25.06
Education (x_2)	3.00	2.00	5.00	4.09	0.82	0.67	19.96
Functional education (x_3)	13.00	2.00	15.00	8.27	3.39	11.52	41.06
Size of family (x_4)	6.00	3.00	9.00	5.62	1.48	2.19	26.34
Cultivated Land Holding per capita (x_5)	1.67	0.25	1.92	0.77	0.30	0.09	38.83
No. of land fragment (x_6)	1.00	1.00	2.00	1.24	0.43	0.19	34.73
Avg. size of land fragment (x_7)	1.00	0.25	1.25	0.64	0.19	0.04	30.18
Distance between land fragment (x_8)	0.50	1.00	1.50	1.26	0.16	0.03	12.73
Homestead land per capita (x_9)	1.57	0.43	2.00	0.81	0.27	0.07	33.78
Marketed Surplus per capita (x_{10})	0.98	0.19	1.17	0.50	0.21	0.04	41.64
On farm income per capita (x_{11})	80316.67	12000.00	92316.67	28521.29	11894.20	141471970.91	41.70
Family Expenditure per capita (x_{12})	58806.89	8668.75	67475.64	19876.87	8770.29	76918047.86	44.12
Savings per capita (x_{13})	12590.12	254.32	12844.44	1778.44	1728.90	2989087.60	97.21
Land under irrigation per capita (x_{14})	1.68	0.08	1.77	0.64	0.31	0.10	47.82
Cropping intensity (x_{15})	296.69	103.31	400.00	128.69	3.99	1431.49	29.40
Livestock count (x_{16})	13.00	2.00	15.00	7.30	2.42	5.88	33.22
Communication variable (x_{17})	24.00	96.00	120.00	105.20	9.36	87.53	8.89
LPG Consumption per person (x_{18})	49.56	15.44	65.00	33.26	13.09	171.32	39.35
Production of organic manure (x_{19})	18.62	3.29	21.90	10.53	4.10	16.83	38.95
Application of organic manure (x_{20})	18.62	5.29	23.90	12.20	4.16	17.32	34.11
Rice stubble height (x_{21})	9.00	2.00	11.00	4.59	2.66	7.08	57.97
Crop residue left in field (x_{22})	8.33	1.25	9.58	4.40	1.59	2.54	36.22
Soil conservation perception (x_{23})	6.50	3.00	9.50	5.58	1.59	2.53	28.49
Water conservation perception (x_{24})	4.50	7.00	11.50	8.80	1.13	1.29	12.90
Bio-diversity conservation perception (x_{25})	4.50	3.00	7.50	4.72	1.03	1.06	21.78
Appl. of climate resilient technology (x_{26})	4.50	1.00	5.50	2.72	1.02	1.04	37.44
Perception on Climate Change (y_1)	5.50	2.00	7.50	4.22	1.24	1.54	29.44
Perceived Uncertainty (y_2)	5.96	3.15	9.12	6.03	1.34	1.81	22.29
Change Dynamics (y_3)	4.10	5.70	9.80	7.98	1.04	1.08	13.02

Result: The Table 13 depicts the distributive characteristics of the variables for the 90 respondents under the study.

Revelation: The pattern of distribution of the variable, Age (x_1) depicts that the maximum age of the respondents is 70 and the minimum is 26. The mean value of age of the respondents is 42.98 with standard deviation of 10.77. The distribution of the variable is strongly consistent with a CV value of 25.06 per cent.

The pattern of distribution of the variable, Education (x_2) depicts that the maximum education of the respondents is 5 and the minimum is 2. The mean value of education of the respondents is 4.09 with standard deviation of 0.82. The Coefficient of Variation is 19.96 per cent which indicates the distribution of variable is highly consistent.

The pattern of distribution of the variable, Functional education (x_3) depicts that the maximum functional education of the respondents is 15 and the minimum is 2. The mean value of functional education of the respondents is 8.27 with standard deviation of 3.39. The distribution of the variable is fairly consistent with a CV value of 41.06 per cent.

The pattern of distribution of the variable, Size of family (x_4) depicts that the maximum size of family of the respondents is 9 and the minimum is 3. The mean value of size of family of the respondents is 5.62 with standard deviation of 1.48. The Coefficient of Variation is 26.34 per cent which indicates the distribution of variable is fairly consistent.

The pattern of distribution of the variable, Cultivated Land Holding per capita (x_5) depicts that the maximum cultivated land holding of the respondents is 1.92 and the minimum is 0.25. The mean value of cultivated land holding of the respondents is 0.77 with standard deviation of 0.30. The distribution of the variable is fairly consistent with a CV value of 38.83 per cent.

The pattern of distribution of the variable, No. of land fragment (x_6) depicts that the maximum no. of land fragment of the respondents is 2 and the minimum is 1. The mean value of no. of land fragment of the respondents is 1.24 with standard deviation of 0.43. The Coefficient of Variation is 34.73 per cent which indicates the distribution of variable is fairly consistent.

The pattern of distribution of the variable, Avg. size of land fragment (x_7) depicts that the maximum avg. size of land fragment of the respondents is 1.25 and the minimum is 0.25. The mean value of avg. size of land fragment of the respondents is 0.64 with standard deviation of 0.19. The distribution of the variable is fairly consistent with a CV value of 30.18 per cent.

The pattern of distribution of the variable, Distance between land fragments (x_8) depicts that the maximum distance between land fragments is 1.50 and the minimum is 1. The mean value of distance between land fragments is 1.26 with standard deviation of 0.16. The Coefficient of Variation is 12.73 per cent which indicates the distribution of variable is highly consistent.

The pattern of distribution of the variable, Homestead land per capita (x_9) depicts that the maximum homestead land of the respondents is 2 and the minimum is 0.43. The mean value of homestead land of the respondents is 0.81 with standard deviation of 0.27. The distribution of the variable is fairly consistent with a CV value of 33.78 per cent.

The pattern of distribution of the variable, Marketed Surplus per capita (x_{10}) depicts that the maximum marketed surplus of the respondents is 1.17 and the minimum is 0.19. The mean value of marketed surplus of the respondents is 0.50 with standard deviation of 0.21. The Coefficient of Variation is 41.64 per cent which indicates the distribution of variable is fairly consistent.

The pattern of distribution of the variable, On farm income per capita (x_{11}) depicts that the maximum on farm income of the respondents is 92316.67 and the minimum is 12000. The mean value of on farm income of the respondents is 28521.29 with standard deviation of 11894.20. The distribution of the variable is fairly consistent with a CV value of 41.70 per cent.

The pattern of distribution of the variable, Family Expenditure per capita (x_{12}) depicts that the maximum family expenditure of the respondents is 67475.64 and the minimum is 8668.75. The mean value of family expenditure of the respondents is 19876.87 with standard deviation of 8770.29. The Coefficient of Variation is 44.12 per cent which indicates the distribution of variable is fairly consistent.

The pattern of distribution of the variable, Savings per capita (x_{13}) depicts that the maximum savings of the respondents is 12844.44 and the minimum is 254.32. The mean value of savings of the respondents is 1778.44 with standard deviation of 1728.90. The distribution of the variable is highly inconsistent with a CV value of 97.21 per cent.

The pattern of distribution of the variable, Land under irrigation per capita (x_{14}) depicts that the maximum land under irrigation is 1.77 and the minimum is 0.08. The mean value of land under irrigation is 0.64 with standard deviation of 0.31. The Coefficient of Variation is 47.82 per cent which indicates the distribution of variable is fairly consistent.

The pattern of distribution of the variable, Cropping intensity (x_{15}) depicts that the maximum cropping intensity is 96.80 and the minimum is 25. The mean value of cropping intensity is 81.31 with standard deviation of 13.78. The distribution of the variable is highly consistent with a CV value of 16.95 per cent.

The pattern of distribution of the variable, Livestock count (x_{16}) depicts that the maximum livestock count is 15 and the minimum is 2. The mean value of livestock count is 7.30 with standard deviation of 2.42. The Coefficient of Variation is 33.22 per cent which indicates the distribution of variable is fairly consistent.

The pattern of distribution of the variable, Communication variable (x_{17}) depicts that the maximum communication variable is 120 and the minimum is 96. The mean value of communication variable is 105.20 with standard deviation of 9.36. The distribution of the variable is very highly consistent with a CV value of 8.89 per cent.

The pattern of distribution of the variable, LPG Consumption per person (x_{18}) depicts that the maximum LPG consumption is 65 and the minimum is 15.44. The mean value of LPG consumption is 33.26 with standard deviation of 13.09. The Coefficient of Variation is 39.35 per cent which indicates the distribution of variable is fairly consistent.

The pattern of distribution of the variable, Production of organic manure (x_{19}) depicts that the maximum organic manureproduction is 21.90 and the minimum is 3.29. The mean value of organic manure production is 10.53 with standard deviation of 4.10. The distribution of the variable is fairly consistent with a CV value of 38.95 per cent.

The pattern of distribution of the variable, Application of organic manure (x_{20}) depicts that the maximum organic manure application is 23.90 and the minimum is 5.29. The mean value of organic manure application is 12.20 with standard deviation of 4.16. The Coefficient of Variation is 34.11 per cent which indicates the distribution of variable is fairly consistent.

The pattern of distribution of the variable, Rice stubble height (x_{21}) depicts that the maximum rice stubble height is 11 and the minimum is 2. The mean value of rice stubble height is 4.59 with standard deviation of 2.66. The distribution of the variable is inconsistent with a CV value of 57.97 per cent.

The pattern of distribution of the variable, Crop residue left in field (x_{22}) depicts that the maximum crop residue left in field is 9.58 and the minimum is 1.25. The mean value of crop residue left in field is 4.40 with standard deviation of 1.59. The Coefficient of Variation is 36.22 per cent which indicates the distribution of variable is fairly consistent.

The pattern of distribution of the variable, Soil conservation perception (x_{23}) depicts that the maximum soil conservation perception is 9.50 and the minimum is 3.00. The mean value of soil conservation perception is 5.58 with standard deviation of 1.59. The distribution of the variable is fairly consistent with a CV value of 28.49 per cent.

The pattern of distribution of the variable, Water conservation perception (x_{24}) depicts that the maximum water conservation perception is 11.50 and the minimum is 7.00. The mean value of water conservation perception is 8.80 with standard deviation of 1.13. The Coefficient of Variation is 12.90 per cent which indicates the distribution of variable is highly consistent.

The pattern of distribution of the variable, Bio-diversity conservation perception (x_{25}) depicts that the maximum bio-diversity conservation perception is 7.50 and the minimum is 3.00. The mean value of bio-diversity conservation perception is 4.72 with standard deviation of 1.03. The distribution of the variable is highly consistent with a CV value of 21.78 per cent.

The pattern of distribution of the variable, Appl. of climate resilient technology (x_{26}) depicts that the maximum climate resilient technology application is 5.50 and the minimum is 1.00. The mean value of climate resilient technology application is 2.72 with standard deviation of 1.02. The Coefficient of Variation is 37.44 per cent which indicates the distribution of variable is fairly consistent.

The pattern of distribution of the variable, Perception on Climate Change (y_1) depicts that the maximum climate change perception is 7.50 and the minimum is 2.00. The mean value of climate change perception is 4.22 with standard deviation of 1.24. The distribution of the variable is fairly consistent with a CV value of 29.44 per cent.

The pattern of distribution of the variable, Perceived Uncertainty (y_2) depicts that the maximum perceived uncertainty is 9.12 and the minimum is 3.15. The mean value of perceived uncertainty is 6.03 with standard deviation of 1.34. The Coefficient of Variation is 22.29 per cent which indicates the distribution of variable is fairly consistent.

The pattern of distribution of the variable, Change Dynamics (y_3) depicts that the maximum change dynamics is 9.80 and the minimum is 5.70. The mean value of change dynamics is 7.98 with standard deviation of 1.04. The distribution of the variable is highly consistent with a CV value of 13.02 per cent.

6.2 Coefficient of Correlation (r): Perception on Climate Change (y_1) vs. 26 Independent Variables ($x_1 - x_{26}$)

Table 14: Coefficient of Correlation (r): y_1 vs. $x_1 - x_{26}$

Sl.No.	Independent Variables	'r' Value	Remarks
1	Age (x_1)	-0.276	**
2	Education (x_2)	-0.117	
3	Functional education (x_3)	0.180	
4	Size of family (x_4)	-0.286	**
5	Cultivated Land Holding per capita (x_5)	0.268	*
6	No. of land fragment (x_6)	0.038	
7	Avg. size of land fragment (x_7)	0.308	**
8	Distance between land fragment (x_8)	-0.097	
9	Homestead land per capita (x_9)	0.016	
10	Marketed Surplus per capita (x_{10})	0.230	*
11	On farm income per capita (x_{11})	0.233	*
12	Family Expenditure per capita (x_{12})	0.303	**
13	Savings per capita (x_{13})	0.059	
14	Land under irrigation per capita (x_{14})	0.249	*
15	Cropping intensity (x_{15})	0.102	
16	Livestock count (x_{16})	0.194	

Sl.No.	Independent Variables	'r' Value	Remarks
17	Communication variable (x_{17})	0.146	
18	LPG Consumption per person (x_{18})	0.302	**
19	Production of organic manure (x_{19})	0.011	
20	Application of organic manure (x_{20})	0.026	
21	Rice stubble height (x_{21})	0.059	
22	Crop residue left in field (x_{22})	0.239	*
23	Soil conservation perception (x_{23})	0.229	*
24	Water conservation perception (x_{24})	0.162	
25	Bio-diversity conservation perception (x_{25})	0.394	**
26	Appl. of climate resilient technology (x_{26})	0.437	**

**Correlation is significant at the 0.01 level

*Correlation is significant at the 0.05 level

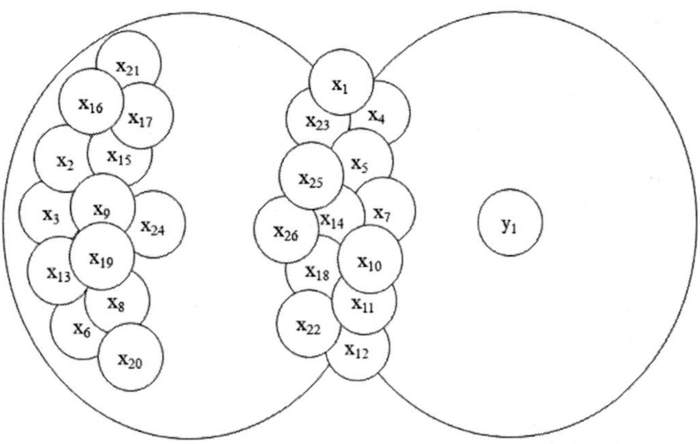

x_1 – Age
x_4 - Size of family
x_7 - Avg. size of land fragment
x_{10} - Marketed Surplus per capita
x_{13} - Savings per capita
x_{16} - Livestock count
x_{19} - Production of organic manure
x_{22} - Crop residue left in field
x_{25} - Bio-diversity conservation perception
x_2 - Education
x_5 - Cultivated Land Holding per capita
x_8 - Distance between land fragment
x_{11} - On farm income per capita
x_{14} - Land under irrigation per capita
x_{17} - Communication variable
x_{20} - Application of organic manure
x_{23} - Soil conservation perception
x_{26} - Application of climate resilient technology
x_3 - Functional education
x_6 - No. of land fragment
x_9 - Homestead land per capita
x_{12} - Family Expenditure per capita
x_{15} - Cropping intensity
x_{18} - LPG Consumption per person
x_{21} - Rice stubble height
x_{24} - Water conservation perception
y_1 - Perception on Climate Change

Coefficient of Correlation (r): Perception on Climate Change (y_1) vs. 26 Independent Variables (x_1 –x_{26})

Model 1: Coefficient of Correlation (r): y_1 vs. $x_1 - x_{26}$

Result: The Table 14 presents the coefficient of correlation (r) between Perception on Climate Change (y_1) and the 26 independent variable. It has been found that 13 independent variables have recorded significant correlation either positive or negative with the dependent variable y_1.

Revelation: It has been found that Age (x_1) and Size of family (x_4) have recorded significantly negative correlation with Perception on Climate Change (y_1). So it suggests that the climate change perception is better as for the younger respondents and respondents hailing from smaller family size background. Through the present primary and secondary education, the young minds are trained up with the concept of climate change and environmental dynamics.

The table also depicts that having higher size of Cultivated Land Holding per capita (x_5) is positively correlated with the climate change perception.

It is interesting to note that the owner of small and fragmented holding have become more vulnerable to climate change while respondents with higher size of land fragments have scope to adapt to the climate change.

Respondents with higher Marketed Surplus per capita (x_{10}), On farm income per capita (x_{11}) and Family Expenditure per capita (x_{12}) have developed better climate change perception. So, for the poorer farming diaspora, the implication of climate change is not that big.

It is also well discernible that farmers having higher size of land holding under irrigation as well as consuming higher volume of LPG have recorded better responses to climate change phenomena vis-a-vis perception.

Crop residue left in field (x_{22}), Soil conservation perception (x_{23}) and Bio-diversity conservation perception (x_{25}) have recorded significantly positive correlation with the climate change perception. When crop residues are kept in the field, they admix with the soil at a higher proportion and the soil retains more organic matter and becomes ecologically resilient. Thus soil conservation is practiced followed by the bio-diversity conservation through conserving the beneficial microbes present in the soil by providing the organic matter at a higher proportion. Thus, all these variables have got mutual, regenerative and reiterative effect on climate mitigation and perception.

Appl. of climate resilient technology (x_{26}) shows significantly positive correlation with the Perception on Climate Change (y_1). The application of climate resilient technology such as minimum and zero tillage, mulching, reduced fertilizer application, sprinkler and drip irrigation, precision agriculture etc. are very much crucial for climate change mitigation and adaptation. So, it can be suggested that higher the uses of these climate resilient technology, better the responses to climate change phenomena vis-a-vis perception.

6.3 Multiple Regression Analysis: Perception on Climate Change (y_1) vs. 26 Causal Variables ($x_1 - x_{26}$)

Table 15: Multiple Regression Analysis: y_1 vs. $x_1 - x_{26}$

Sl.No.	Variables	Reg. Coef. B	S. E. B	Beta	t Value
1	Age (x_1)	-0.018	0.022	-0.158	-0.843
2	Education (x_2)	-0.453	0.398	-0.297	-1.137
3	Functional education (x_3)	0.100	0.088	0.274	1.138
4	Size of family (x_4)	0.181	0.288	0.216	0.627

5	Cultivated Land Holding per capita (x_5)	7.469	6.777	1.803	1.102
6	No. of land fragment (x_6)	0.246	1.677	0.085	0.146
7	Avg. size of land fragment (x_7)	0.980	3.694	0.152	0.265
8	Distance between land fragment (x_8)	-0.225	0.820	-0.029	-0.274
9	Homestead land per capita (x_9)	-1.782	0.752	-0.391	-2.371
10	Marketed Surplus per capita (x_{10})	0.380	0.837	0.064	0.454
11	On farm income per capita (x_{11})	7.98E-05	0.0001	0.764	0.794
12	Family Expenditure per capita (x_{12})	-9.24E-05	0.0001	-0.653	-0.888
13	Savings per capita (x_{13})	0.0004	0.0004	-0.647	-1.124
14	Land under irrigation per capita (x_{14})	-4.223	4.958	-1.048	-0.852
15	Cropping intensity (x_{15})	0.025	0.029	0.280	0.867
16	Livestock count (x_{16})	-0.214	0.204	-0.418	-1.049
17	Communication variable (x_{17})	-0.001	0.015	-0.009	-0.077
18	LPG Consumption per person (x_{18})	0.016	0.021	0.169	0.762
19	Production of organic manure (x_{19})	-0.112	0.499	-0.368	-0.223
20	Application of organic manure (x_{20})	0.063	0.474	0.211	0.133
21	Rice stubble height (x_{21})	0.078	0.089	0.166	0.869
22	Crop residue left in field (x_{22})	-0.158	0.338	-0.203	-0.468
23	Soil conservation perception (x_{23})	0.108	0.095	0.139	1.142
24	Water conservation perception (x_{24})	0.054	0.127	0.049	0.421
25	Bio-diversity conservation perception (x_{25})	0.113	0.148	0.093	0.765
26	Appl. of climate resilient technology (x_{26})	0.407	0.176	0.334	2.317

R Square value: ***47.4%***

The Standard Error of the Estimate: ***1.07048***

Result: The Table 15 presents the full model of multiple regression analysis where the Beta value implies the unit of functional impact to generate consequential impact on the dependent variable.

Revelation: The R^2value being 47.4 per cent, it can be concluded that a combination of 26 causal variables has contributed 47.4 per cent of variance embedded with y_1 and these 26 causal variables have got a certain level of proficiency for generating causal impact on the dependent variable y_1.

It has been found that Cultivated Land Holding per capita (x_5) has the highest beta value. It implies that this causal variable has got the highest substantive effect on the Perception on Climate Change (y_1) and emerged as the strongest functional determinant of y_1.

6.4 Stepwise Regression Analysis: Perception on Climate Change (y_1) vs. 26 Causal Variables ($x_1 - x_{26}$)

Table 16: Stepwise Regression Analysis: y_1 vs. $x_1 - x_{26}$

Sl. No	Variables	Reg. Coef. B	S.E. B	Beta	t value
1	Appl. of climate resilient technology (x26)	0.454	0.116	0.372	3.924
2	Bio-diversity conservation perception (x25)	0.236	0.118	0.195	1.998
3	Avg. size of land fragment (x7)	2.110	0.696	0.327	3.031
4	Homestead land per capita (x9)	-1.005	0.481	-0.220	-2.088

R square: ***33.7%***

The standard error of the estimate: ***1.0348***

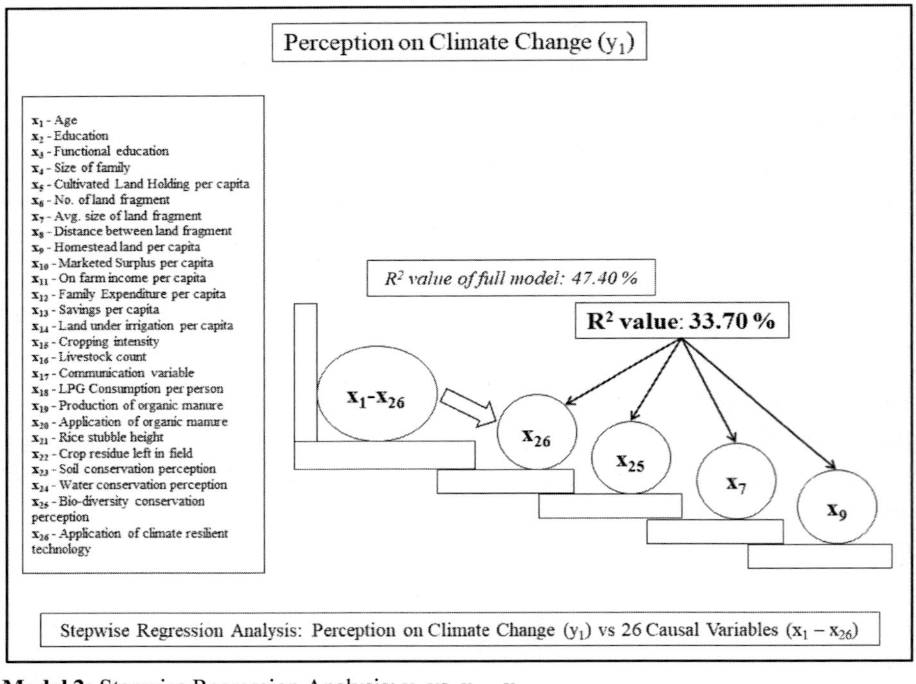

Model 2: Stepwise Regression Analysis: y_1 vs. $x_1 - x_{26}$

Considering the results of full model of multiple regression analysis, stepwise regression analysis has been persuaded to extract the dominant variables with strong functional impact on y_1.

Result: The Table 16 reveals that through the stepwise elimination, the following variables are retained at the last stage of the analysis. These are Appl. of climate resilient technology (x_{26}), Bio-diversity conservation perception (x_{25}), Avg. size of land fragment (x_7) and Homestead land per capita (x_9).

Revelation: It is well discernible that the strongest determinant in characterizing and functionally impacting the Perception on climate change (y_1) has been the Appl. of climate resilient technology (x_{26}) with the highest beta value.

Climate resilient technology like water saving, mulching, minimum/zero tillage and reduced fertilizer application etc. has generated a binary of perception on climate change at a large through the better perception on climate change and perceived ecological services and resilience by the farmer.

R^2value being 33.7 per cent, it is to infer that a cumulative contribution of these four variables to the variation in y_1 stands at 71.09 per cent. It is well discernible that these four variables together have contributed in 71.09 per cent of total variance explained. These four variables have got tremendous strategic and functional implications.

6.5 Path Analysis: Decomposition of Total Effect into Direct, Indirect and Residual Effect: Perception on Climate Change (y_1)

Table 17: Path Analysis: y_1 vs. $x_1 - x_{26}$

Sl. No	Variables	Total Effect	Direct Effect	Indirect Effect	Highest Indirect Effect
1	Age (x_1)	-0.276	-0.149	-0.127	-0.825 (x_5)
2	Education (x_2)	-0.117	-0.294	0.177	-0.299 (x_{14})
3	Functional education (x_3)	0.180	0.274	-0.094	0.763 (x_5)
4	Size of family (x_4)	-0.286	0.207	-0.493	0.601 (x_{14})
5	Cultivated Land Holding per capita (x_5)	0.268	1.820	-1.552	-1.045 (x_{14})
6	No. of land fragment (x_6)	0.038	0.081	-0.043	1.152 (x_5)
7	Avg. size of land fragment (x_7)	0.308	0.148	0.160	1.003 (x_5)
8	Distance between land fragment (x_8)	-0.097	-0.028	-0.069	-0.073 (x_{13})
9	Homestead land per capita (x_9)	0.016	-0.393	0.409	0.930 (x_5)
10	Marketed Surplus per capita (x_{10})	0.230	0.066	0.164	1.085 (x_5)
11	On farm income per capita (x_{11})	0.233	0.740	-0.507	1.269 (x_5)
12	Family Expenditure per capita (x_{12})	0.303	-0.639	0.942	1.028 (x_5)
13	Savings per capita (x_{13})	0.059	-0.637	0.696	1.183 (x_5)
14	Land under irrigation per capita (x_{14})	0.249	-1.078	1.327	1.764 (x_5)
15	Cropping intensity (x_{15})	0.102	0.282	-0.180	0.737 (x_5)
16	Livestock count (x_{16})	0.194	-0.418	0.612	1.261 (x_5)
17	Communication variable (x_{17})	0.146	-0.007	0.153	-0.231 (x_5)
18	LPG Consumption per person (x_{18})	0.302	0.153	0.149	0.768 (x_5)
19	Production of organic manure (x_{19})	0.011	-0.521	0.531	1.083 (x_5)
20	Application of organic manure (x_{20})	0.026	0.346	-0.320	1.099 (x_5)
21	Rice stubble height (x_{21})	0.059	0.174	-0.115	1.119 (x_5)
22	Crop residue left in field (x_{22})	0.239	-0.172	0.411	1.689 (x_5)
23	Soil conservation perception (x_{23})	0.229	0.144	0.085	0.121 (x_2)

Sl. No	Variables	Total Effect	Direct Effect	Indirect Effect	Highest Indirect Effect
24	Water conservation perception (x_{24})	0.162	0.048	0.114	-0.131 (x_5)
25	Bio-diversity conservation perception (x_{25})	0.394	0.093	0.301	0.344 (x_5)
26	Appl. of climate resilient technology (x_{26})	0.437	0.333	0.104	0.228 (x_5)

Residual effect: *0.52544*

Highest individual indirect effect: *Cultivated Land Holding per capita (x_5)* [Frequency: 21]

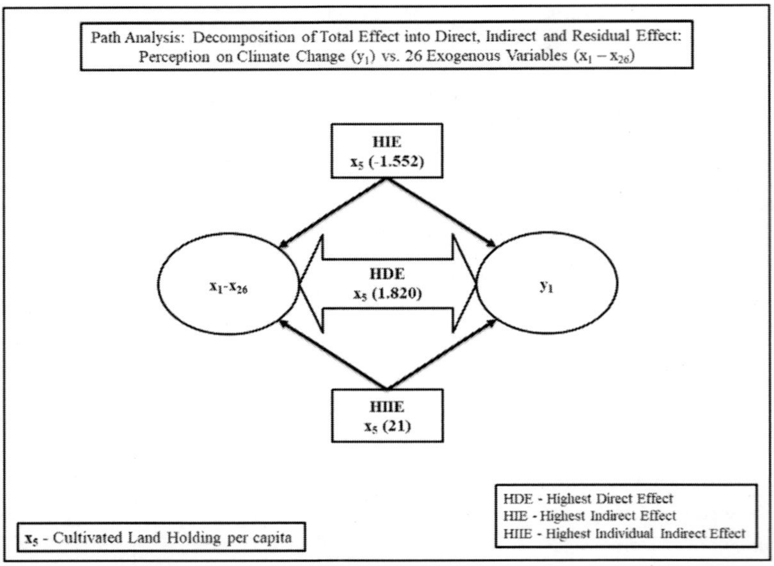

Model 3: Path Analysis: y_1 vs. $x_1 - x_{26}$

Result: Table 17presents the path analysis which has decomposed the overall impact of exogenous variables on the consequent variable into direct, indirect, and residual effects.

Revelation: The exogenous variable Cultivated Land Holding per capita (x_5) has recorded the highest direct effect on the consequent variable Perception on Climate Change (y_1). It certainty concludes that cultivated land holding has been the strongest predictor for estimating the climate change perception.

The variable has also exerted the highest indirect effect on the consequent variable by involving as many as 25 other exogenous variables. So, this variable has got significant strategic importance.

The exogenous variable, Appl. of climate resilient technology (x_{26}) has been found to exert the highest total effect on the consequent variable Perception on Climate Change (y_1).

The residual effect being 52.5 percent, it is to conclude that with a combination of 26 exogenous variables, 52.5 percent of variance in the consequent variable, Perception on Climate Change (y_1) could not be explained.

6.6 Coefficient of Correlation (r): Perceived Uncertainty (y_2) vs. 26 Independent Variables ($x_1 - x_{26}$)

Table 18: Coefficient of Correlation (r): y_2 vs. $x_1 - x_{26}$

Sl. No.	Independent Variables	'r' Value	Remarks
1	Age (x_1)	-0.319	**
2	Education (x_2)	-0.045	
3	Functional education (x_3)	0.225	*
4	Size of family (x_4)	-0.477	**
5	Cultivated Land Holding per capita (x_5)	0.197	
6	No. of land fragment (x_6)	-0.169	
7	Avg. size of land fragment (x_7)	0.421	**
8	Distance between land fragment (x_8)	-0.173	
9	Homestead land per capita (x_9)	0.270	*
10	Marketed Surplus per capita (x_{10})	0.155	
11	On farm income per capita (x_{11})	0.208	*
12	Family Expenditure per capita (x_{12})	0.266	*
13	Savings per capita (x_{13})	0.141	
14	Land under irrigation per capita (x_{14})	0.188	
15	Cropping intensity (x_{15})	0.095	
16	Livestock count (x_{16})	-0.094	
17	Communication variable (x_{17})	0.267	*
18	LPG Consumption per person (x_{18})	0.441	**
19	Production of organic manure (x_{19})	-0.280	**
20	Application of organic manure (x_{20})	-0.256	*
21	Rice stubble height (x_{21})	-0.119	
22	Crop residue left in field (x_{22})	0.113	
23	Soil conservation perception (x_{23})	0.302	**
24	Water conservation perception (x_{24})	0.281	**
25	Bio-diversity conservation perception (x_{25})	0.378	**
26	Appl. of climate resilient technology (x_{26})	0.265	*

**Correlation is significant at the 0.01 level

*Correlation is significant at the 0.05 level

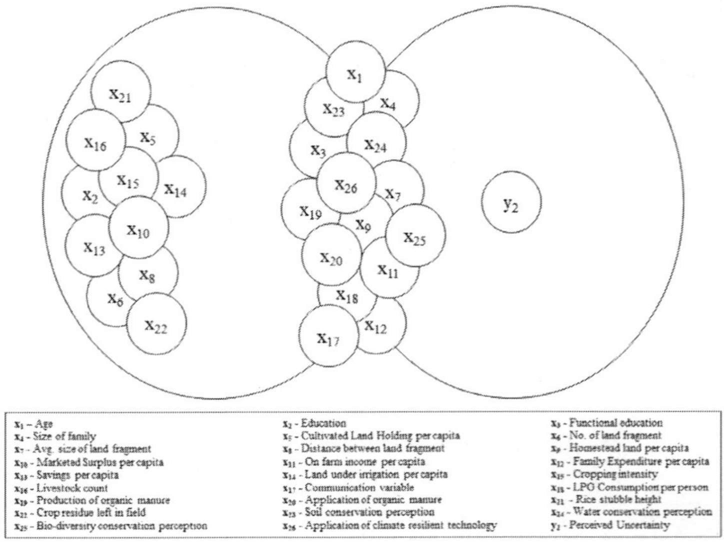

Coefficient of Correlation (r): Perceived Uncertainty (y_2) vs. 26 Independent Variables ($x_1 - x_{26}$)

Model 4: Coefficient of Correlation (r): y_2 vs. $x_1 - x_{26}$

Result: The Table 18 presents the coefficient of correlation (r) between Perceived Uncertainty (y_2) and the 26 independent variable. It has been found that 15 independent variables have recorded significant correlation either positive or negative with the dependent variable y_2.

Revelation: It has been found that Age (x_1) has recorded significantly negative correlation with the dependent variable. The farmers, aged and experienced by default has confronted more with uncertainty, risk and vulnerability. So, they are more experienced and exposed to uncertainty.

The independent variable Functional education (x_3) has been found to be positively correlated with the dependent variable at 5% significance level. It suggests that with more no. of training, visit to research institutes and interaction with scientist etc. confidence and capacity builds amongst the farmers and they tend to be innovative to some extent by taking up latest technologies in practice. This may expose them to higher level of uncertainty.

It is elicited that respondents hailing from small family size background are more vulnerable to risk. The smaller is the family size, the higher has been the risk loads and inflicts.

It has been found that Avg. size of land fragment (x_7) and Homestead land per capita (x_9) have recorded significantly positive correlation with the dependent variable y_2. It suggests that management of higher size of land unit leads to an increase in perceived uncertainty.

Result shows that both On farm income per capita (x_{11}) and Family Expenditure per capita (x_{12}) arepositively correlated with the dependent variable y_2 at 1% significance level. It portrays that better returns from farms leads to perceive the uncertainty more and subsequently expenditure becomes more in response to the perceived uncertainty.

It is elicited that ineffective communication makes farmers ignorant and uninformed about latest technologies or innovations, trainings, insurances or schemes etc. which are supposed to improve their capacity, farming activities, production, profit and ultimately livelihood. Thus eventually makes them highly uncertain on how to response better to the fast changing world as well as to the uncertain market and weather.

It is elicited that respondents consuming higher volume of LPG have perceived uncertainty more. It suggests that usage of LPG leads to higher family expenditure and in case of small and marginal farmers, it eventually leads to the more exposure to the uncertainty.

It has been found that Production of organic manure (x_{19}) and Application of organic manure (x_{20}) are negatively correlated with Perceived Uncertainty (y_2) at 1% and 5% level of significance respectively. As more production of organic manure and application of that manure afterwards in the crop field helps in better soil and bio-diversity management by means of increasing overall soil quality in long turn, it can be inferred that productivity followed by profit will increase also and the exposure to uncertainty will decrease eventually.

Soil conservation perception (x_{23}), Water conservation perception (x_{24}), Bio-diversity conservation perception (x_{25}) and Appl. of climate resilient technology (x_{26}) have been found positively correlated with Perceived Uncertainty (y_2). It can be inferred that better perception on managing soil, water, bio-diversity and more utilization of climate resilient technology comes from more exposure to the brunt of climate change and uncertainty generates thereafter.

6.7 Multiple Regression Analysis: Perceived Uncertainty (y_2) vs. 26 Causal Variables ($x_1 - x_{26}$)

Table 19: Multiple Regression Analysis: y_2 vs. $x_1 - x_{26}$

Sl.No.	Variables	Reg. Coef. B	S. E. B	Beta	t Value
1	Age (x_1)	0.035	0.020	0.284	1.782
2	Education (x_2)	-0.231	0.366	-0.14	-0.631
3	Functional education (x_3)	0.189	0.081	0.478	2.337
4	Size of family (x_4)	0.018	0.265	0.02	0.069
5	Cultivated Land Holding per capita (x_5)	13.219	6.236	**2.948**	2.120
6	No. of land fragment (x_6)	1.104	1.543	0.355	0.716

Sl.No.	Variables	Reg. Coef. B	S. E. B	Beta	t Value
7	Avg. size of land fragment (x_7)	2.282	3.399	0.327	0.671
8	Distance between land fragment (x_8)	-1.999	0.755	-0.239	-2.648
9	Homestead land per capita (x_9)	-1.121	0.691	-0.227	-1.622
10	Marketed Surplus per capita (x_{10})	-0.178	0.77	-0.028	-0.232
11	On farm income per capita (x_{11})	-3.82E-05	0.00009	-0.338	-0.413
12	Family Expenditure per capita (x_{12})	-1.20E-05	0.00009	-0.078	-0.125
13	Savings per capita (x_{13})	7.19E-05	0.0003	0.093	0.189
14	Land under irrigation per capita (x_{14})	-8.553	4.562	**-1.961**	-1.875
15	Cropping intensity (x_{15})	0.044	0.027	0.452	1.646
16	Livestock count (x_{16})	-0.283	0.188	-0.510	-1.507
17	Communication variable (x_{17})	0.020	0.014	0.136	1.409
18	LPG Consumption per person (x_{18})	-0.030	0.019	-0.293	-1.549
19	Production of organic manure (x_{19})	-0.523	0.46	**-1.597**	-1.139
20	Application of organic manure (x_{20})	0.231	0.436	0.715	0.529
21	Rice stubble height (x_{21})	0.125	0.082	0.247	1.520
22	Crop residue left in field (x_{22})	-0.246	0.311	-0.291	-0.791
23	Soil conservation perception (x_{23})	0.265	0.087	0.313	3.029
24	Water conservation perception (x_{24})	0.211	0.117	0.178	1.807
25	Bio-diversity conservation perception (x_{25})	0.224	0.136	0.172	1.650
26	Appl. of climate resilient technology (x_{26})	0.047	0.162	0.035	0.289

R square: *62%*

The standard error of the estimate: ***0.98494***

Result: The Table 19 presents the full model of multiple regression analysis where the Beta value implies the unit of functional impact to generate consequential impact on the dependent variable.

Revelation: The R^2value being 62 per cent, it can be concluded that a combination of 26 causal variables has contributed 62 per cent of variance embedded with y_2 and these 26 causal variables have got a certain level of proficiency for generating causal impact on the dependent variable y_2.

It has been found that Cultivated Land Holding per capita (x_5) has the highest beta value. It implies that this causal variable has got the highest substantive effect on the Perceived Uncertainty (y_2) and emerged as the strongest functional determinant of y_2.

6.8 Stepwise Regression Analysis: Perceived Uncertainty (y_2) vs. 26 Causal Variables ($x_1 - x_{26}$)

Table 20: Stepwise Regression Analysis: y_2 vs. $x_1 - x_{26}$

Sl.No	Variables	Reg. Coef. B	S. E. B	Beta	t value
1	Size of family (x_4)	-0.342	0.082	-0.377	-4.149
2	Soil conservation perception (x_{23})	0.179	0.074	0.211	2.418
3	Production of organic manure (x_{19})	-0.069	0.028	-0.211	-2.433
4	Bio-diversity conservation perception (x_{25})	0.253	0.120	0.194	2.107

R square: *37.6%*

The standard error of the estimate: *1.0865*

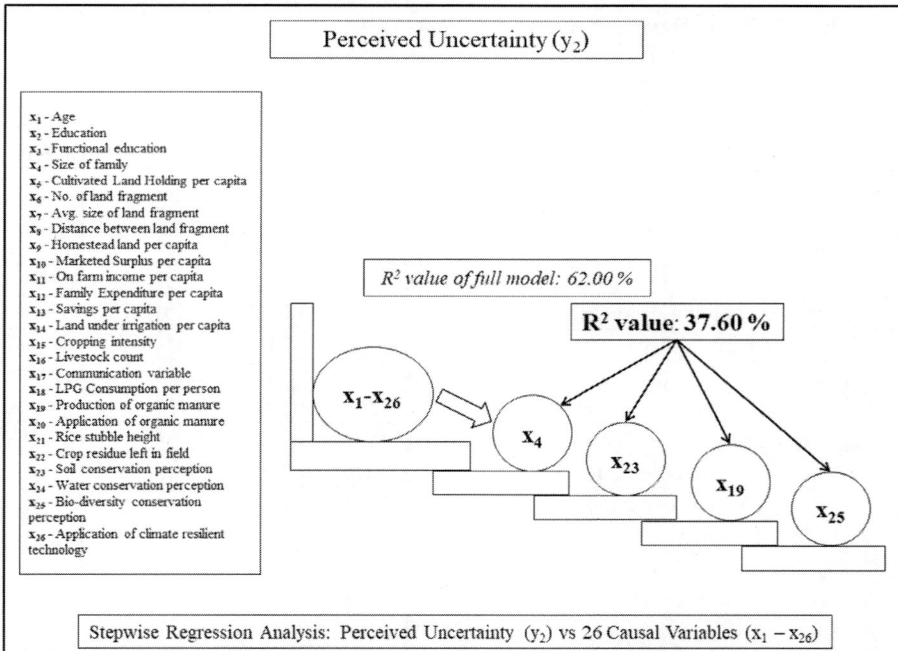

Model 5: Stepwise Regression Analysis: y_2 vs. $x_1 - x_{26}$

Considering the results of full model of multiple regression analysis, stepwise regression analysis has been persuaded to extract the dominant variables with strong functional impact on y_2.

Result: The Table 20 reveals that through the stepwise elimination, the following variables are retained at the last stage of the analysis. These are Size of family (x_4), Soil conservation perception (x_{23}), Production of organic manure (x_{19}) and Bio-diversity conservation perception (x_{25}).

Revelation: It is well discernible that the strongest determinant in characterizing and functionally impacting the Perceived Uncertainty (y_2) has been the Size of family (x_4) with the highest beta value.

Respondents hailing from small family size background are more vulnerable to risk. The smaller is the family size, the higher has been the risk loads and inflicts.

R^2value being 37.6 per cent, it is to infer that a cumulative contribution of these four variables to the variation in y_2 stands at 60.6 per cent. It is well discernible that these four variables together have contributed in 60.6 per cent of total variance explained. These four variables have got tremendous strategic and functional implications.

6.9 Path Analysis: Decomposition of Total Effect into Direct, Indirect and Residual Effect: Perceived Uncertainty (y_2)

Table 21: Path Analysis: y_2 vs. $x_1 - x_{26}$

Sl. No	Variables	Total Effect	Direct Effect	Indirect Effect	Highest Indirect Effect
1	Age (x_1)	-0.319	0.292	-0.611	-1.489 (x_5)
2	Education (x_2)	-0.045	-0.133	0.088	0.796 (x_5)
3	Functional education (x_3)	0.225	0.48	-0.255	1.378 (x_5)
4	Size of family (x_4)	**-0.477**	-0.025	-0.452	-1.894 (x_5)
5	Cultivated Land Holding per capita (x_5)	0.197	**3.288**	**-3.091**	-2.062 (x_{14})
6	No. of land fragment (x_6)	-0.169	0.211	-0.380	2.081 (x_5)
7	Avg. size of land fragment (x_7)	0.421	0.185	0.236	1.812 (x_5)
8	Distance between land fragment (x_8)	-0.173	-0.246	0.073	0.080 (x_3)
9	Homestead land per capita (x_9)	0.270	-0.234	0.504	1.680 (x_5)
10	Marketed Surplus per capita (x_{10})	0.155	-0.007	0.162	1.960 (x_5)
11	On farm income per capita (x_{11})	0.208	-0.348	0.556	2.292 (x_5)
12	Family Expenditure per capita (x_{12})	0.266	-0.094	0.360	1.858 (x_5)
13	Savings per capita (x_{13})	0.141	0.082	0.059	2.137 (x_5)
14	Land under irrigation per capita (x_{14})	0.188	-2.128	2.316	3.186 (x_5)
15	Cropping intensity (x_{15})	0.095	0.51	-0.415	1.332 (x_5)
16	Livestock count (x_{16})	-0.094	-0.455	0.361	2.278 (x_5)
17	Communication variable (x_{17})	0.267	0.136	0.131	-0.418 (x_5)
18	LPG Consumption per person (x_{18})	0.441	-0.285	0.726	1.387 (x_5)
19	Production of organic manure (x_{19})	-0.280	-1.255	0.975	1.956 (x_5)
20	Application of organic manure (x_{20})	-0.256	0.361	-0.617	1.986 (x_5)
21	Rice stubble height (x_{21})	-0.119	0.242	-0.361	2.022 (x_5)
22	Crop residue left in field (x_{22})	0.113	-0.366	0.479	3.051 (x_5)
23	Soil conservation perception (x_{23})	0.302	0.319	-0.017	0.211 (x_{14})
24	Water conservation perception (x_{24})	0.281	0.18	0.101	0.240 (x_{19})
25	Bio-diversity conservation perception (x_{25})	0.378	0.17	0.208	0.621 (x_5)
26	Appl. of climate resilient technology (x_{26})	0.265	0.028	0.237	0.411 (x_5)

Residual effect: ***0.37565***

Highest individual indirect effect: ***Cultivated Land Holding per capita (x_5)*** **[Frequency: 22]**

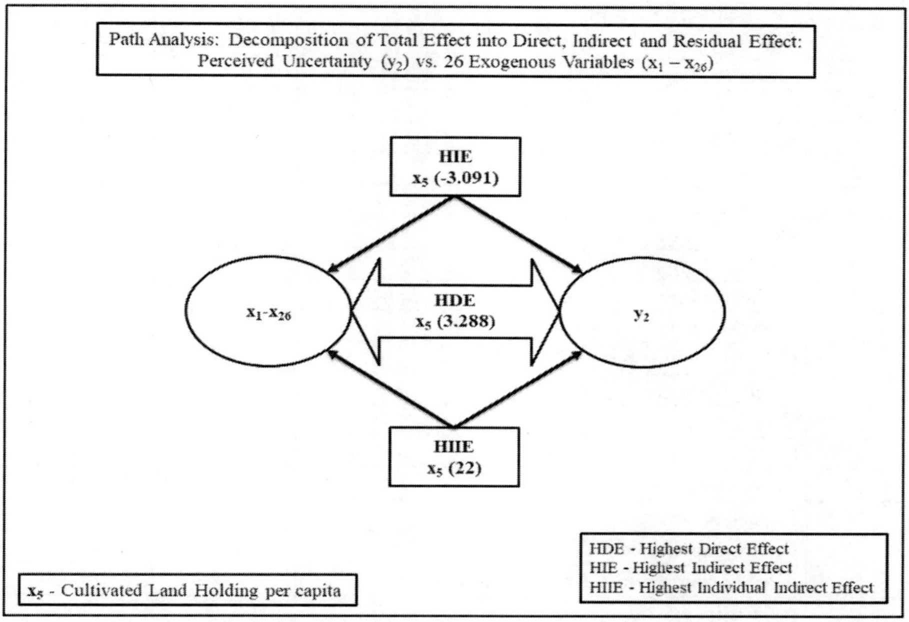

Model 6: Path Analysis: y_2 vs. $x_1 - x_{26}$

Result: Table 21presents the path analysis which has decomposed the overall impact of exogenous variables on the consequent variable into direct, indirect, and residual effects.

Revelation: The exogenous variable Cultivated Land Holding per capita (x_5) has recorded the highest direct effect on the consequent variable Perceived Uncertainty (y_2). It certainly concludes that cultivated land holding has been the strongest predictor for estimating the level of uncertainty.

The variable has also exerted the highest indirect effect on the consequent variable by involving as many as 25 other exogenous variables. So, this variable has got significant strategic importance.

The exogenous variable, Size of family (x_4) has been found to exert the highest total effect on the consequent variable Perceived Uncertainty (y_2).

The residual effect being 37.5 percent, it is to conclude that with a combination of 26 exogenous variables, 37.5 percent of variance in the consequent variable, Perceived Uncertainty (y_2) could not be explained.

6.10 Coefficient of Correlation (r): Change Dynamics (y_3) vs. 26 Independent Variables ($x_1 - x_{26}$)

Table 22: Coefficient of Correlation (r): y_3 vs. $x_1 - x_{26}$

Sl.No.	Independent Variables	'r' Value	Remarks
1	Age (x_1)	0.071	
2	Education (x_2)	0.184	
3	Functional education (x_3)	0.034	
4	Size of family (x_4)	0.021	
5	Cultivated Land Holding per capita (x_5)	-0.125	
6	No. of land fragment (x_6)	-0.036	
7	Avg. size of land fragment (x_7)	-0.118	
8	Distance between land fragment (x_8)	0.204	
9	Homestead land per capita (x_9)	-0.106	
10	Marketed Surplus per capita (x_{10})	0.086	
11	On farm income per capita (x_{11})	-0.104	
12	Family Expenditure per capita (x_{12})	-0.191	
13	Savings per capita (x_{13})	0.023	
14	Land under irrigation per capita (x_{14})	-0.096	
15	Cropping intensity (x_{15})	0.046	
16	Livestock count (x_{16})	-0.305	**
17	Communication variable (x_{17})	-0.379	**
18	LPG Consumption per person (x_{18})	-0.040	
19	Production of organic manure (x_{19})	-0.111	
20	Application of organic manure (x_{20})	-0.130	
21	Rice stubble height (x_{21})	-0.018	
22	Crop residue left in field (x_{22})	-0.063	
23	Soil conservation perception (x_{23})	-0.266	*
24	Water conservation perception (x_{24})	-0.193	
25	Bio-diversity conservation perception (x_{25})	-0.144	
26	Appl. of climate resilient technology (x_{26})	-0.261	*

**Correlation is significant at the 0.01 level

*Correlation is significant at the 0.05 level

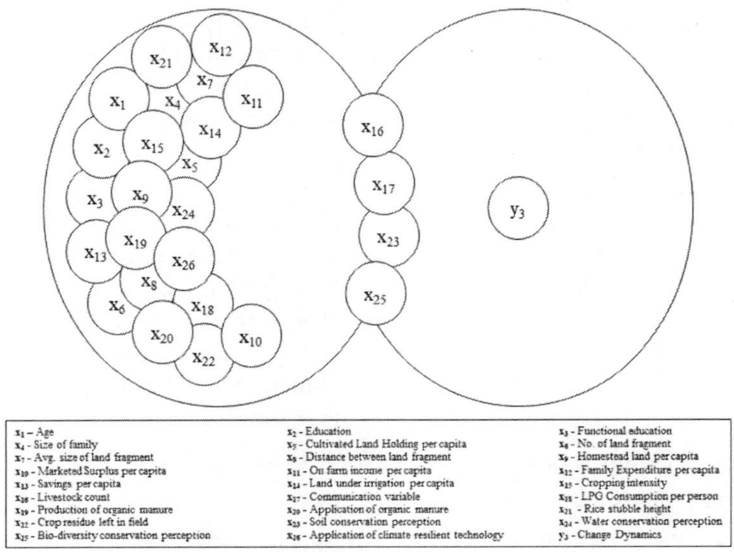

Coefficient of Correlation (r): Change Dynamics (y_3) vs. 26 Independent Variables ($x_1 - x_{26}$)

Model 7: Coefficient of Correlation (r): y_3 vs. $x_1 - x_{26}$

Result: The Table 22 presents the coefficient of correlation (r) between Change Dynamics (y_3) and the 26 independent variable. It has been found that 4 independent variables have recorded significantly negative correlation with the dependent variable y_3.

Revelation: It has been found that Livestock count (x_{16}) has recorded significantly negative correlation with the dependent variable. It depicts the present scenario where livestock are decreasing day by day hence affecting the farm economy in an adverse way. Livestock are the sources of raw material for production of organic manure which being applied in the crop field, conserves the soil and bio-diversity hence increasing the productivity as well as decreases the uses of chemical fertilizers hence assures better return from the farm.

The independent variable, Communication variable (x_{17}) has been found to be negatively correlated with the dependent variable y_3 at 1% significance level. Communication plays a very crucial role in disseminating latest technologies or innovations into the farming community. The technology or innovation ultimately may lead to a better return from the farm. Thus communication should be effective in order to achieve a desired change. However, the statistical result suggests that there is lack of effective communication.

Soil conservation perception (x_{23}) has been found to exert significantly negative correlation with the dependent variable. As a result of high input intensive agriculture practices over the time, soil health, fertility and overall soil quality is deteriorating significantly. However farmers are still practicing high external input agriculture (HEIA) with an expectation to get better return from the farm. Thus the soil conservation aspects has been significantly neglected by the farming community at a large.

Appl. of climate resilient technology (x_{26}) has also been found to exert significantly negative correlation with the dependent variable. As a result of ineffective communication as discussed earlier, climate resilient technology such as minimum and zero tillage, mulching, reduced fertilizer application, sprinkler and drip irrigation, precision agriculture etc. are still not in practice at a large among the farming community of the research locale. However desired change dynamics of a farm can be achieved by means of such climate resilient interventions which already has been demonstrated in several other studies.

6.11 Multiple Regression Analysis: Change Dynamics (y_3) vs. 26 Causal Variables ($x_1 - x_{26}$)

Table 23: Multiple Regression Analysis: y_3 vs. $x_1 - x_{26}$

Sl. No.	Variables	Reg. Coef. B	S. E. B	Beta	t Value
1	Age (x_1)	0.030	0.018	0.316	1.703
2	Education (x_2)	0.044	0.330	0.035	0.135
3	Functional education (x_3)	-0.013	0.073	-0.042	-0.177
4	Size of family (x_4)	0.272	0.239	0.388	1.139
5	Cultivated Land Holding per capita (x_5)	-2.955	5.615	-0.853	-0.526
6	No. of land fragment (x_6)	1.832	1.389	0.762	1.319
7	Avg. size of land fragment (x_7)	4.406	3.061	0.816	1.440
8	Distance between land fragment (x_8)	0.657	0.680	0.102	0.967
9	Homestead land per capita (x_9)	-0.743	0.623	-0.195	-1.193
10	Marketed Surplus per capita (x_{10})	0.775	0.694	0.155	1.118
11	On farm income per capita (x_{11})	0.0001	0.00008	**1.378**	1.446
12	Family Expenditure per capita (x_{12})	0.0001	0.00008	**-1.209**	-1.660
13	Savings per capita (x_{13})	0.0004	0.0003	-0.743	-1.303
14	Land under irrigation per capita (x_{14})	2.218	4.108	0.658	0.540
15	Cropping intensity (x_{15})	-0.008	0.024	-0.108	-0.337
16	Livestock count (x_{16})	-0.526	0.169	**-1.229**	-3.113
17	Communication variable (x_{17})	-0.026	0.012	-0.235	-2.085
18	LPG Consumption per person (x_{18})	0.017	0.017	0.218	0.990
19	Production of organic manure (x_{19})	0.121	0.414	0.476	0.292
20	Application of organic manure (x_{20})	-0.123	0.393	-0.491	-0.312
21	Rice stubble height (x_{21})	0.032	0.074	0.083	0.436
22	Crop residue left in field (x_{22})	0.330	0.280	0.507	1.180

Sl. No.	Variables	Reg. Coef. B	S. E. B	Beta	t Value
23	Soil conservation perception (x_{23})	-0.024	0.079	-0.036	-0.300
24	Water conservation perception (x_{24})	0.011	0.105	0.012	0.103
25	Bio-diversity conservation perception (x_{25})	-0.027	0.122	-0.027	-0.221
26	Appl. of climate resilient technology (x_{26})	0.142	0.146	0.140	0.978

R square: *48.4%*

The standard error of the estimate: *0.88696*

Result: The Table 23 presents the full model of multiple regression analysis where the Beta value implies the unit of functional impact to generate consequential impact on the dependent variable.

Revelation: The R^2value being 48.4 per cent, it can be concluded that a combination of 26 causal variables has contributed 48.4 per cent of variance embedded with y_3 and these 26 causal variables have got a certain level of proficiency for generating causal impact on the dependent variable y_3.

It has been found that On farm income per capita (x_{11}) hasthe highest beta value. It implies that this causal variable has got the highest substantive effect on the Change Dynamics (y_3) and emerged as the strongest functional determinant of y_3.

6.12 Stepwise Regression Analysis: Change Dynamics (y_3) vs. 26 Causal Variables ($x_1 – x_{26}$)

Table 24: Stepwise Regression Analysis: y_3 vs. $x_1 – x_{26}$

Sl. No	Variables	Reg. Coef. B	S. E. B	Beta	t value
1	Communication variable (x_{17})	-0.045	0.010	-0.401	-4.322
2	Livestock count (x_{16})	-0.142	0.040	-0.332	-3.577

R square: 25.4%

The standard error of the estimate: 0.90757

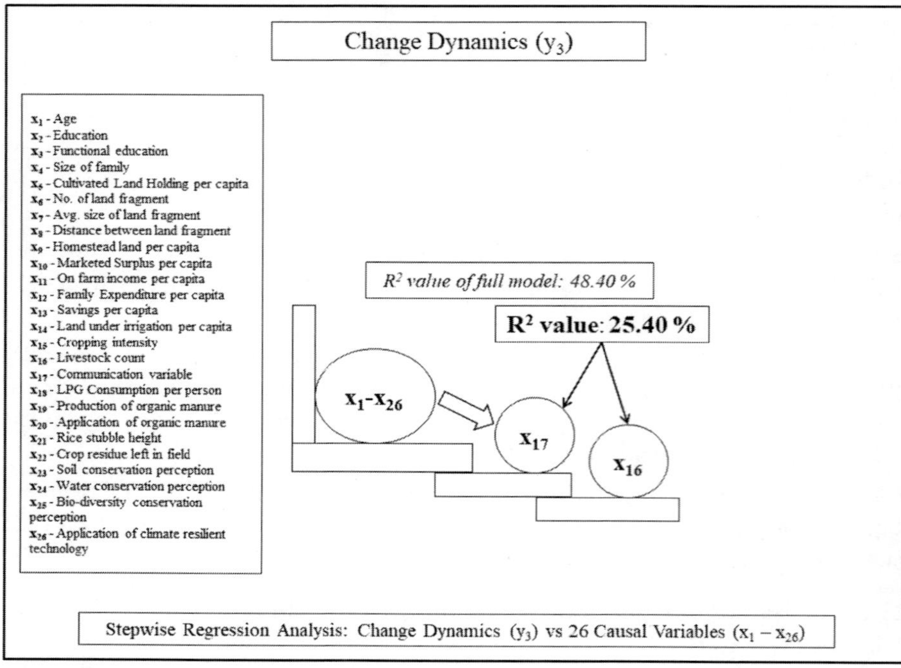

Model 8: Stepwise Regression Analysis: y_3 vs. $x_1 - x_{26}$

Considering the results of full model of multiple regression analysis, stepwise regression analysis has been persuaded to extract the dominant variables with strong functional impact on y_3.

Result: The Table 24 reveals that through the stepwise elimination, the following variables are retained at the last stage of the analysis. These are Communication variable (x_{17}) and Livestock count (x_{16}).

Revelation: It is well discernible that the strongest determinant in characterizing and functionally impacting the Change Dynamics (y_3) has been the Communication variable (x_{17}) with the highest beta value.

Communication by means of different mass media such as newspaper, radio, television etc., opinion leaders, Extension agents and like that is very much essential so to be aware of the latest technologies in practice for agricultural and allied activities, adoption of which eventually leads to the increase in production and/or productivity and subsequently increase in income. However the communication should be effective to the audience. If not it may lead to a reciprocal result instead of the desired one.

R^2value being 25.4 per cent, it is to infer that a cumulative contribution of these two variables to the variation in y_3 stands at 52.4 per cent. It is well discernible that these two variables together have contributed in 52.4 per cent of total variance explained. These two variables have got tremendous strategic and functional implications.

6.13 Path Analysis: Decomposition of Total Effect into Direct, Indirect and Residual Effect: Change Dynamics (y_3)

Table 25: Path Analysis: y_3 vs. $x_1 - x_{26}$

Sl. No	Variables	Total Effect	Direct Effect	Indirect Effect	Highest Indirect Effect
1	Age (x_1)	0.071	0.305	-0.234	-0.569 (x_{11})
2	Education (x_2)	0.184	0.033	0.151	0.337 (x_{11})
3	Functional education (x_3)	0.034	-0.044	0.078	0.651 (x_{11})
4	Size of family (x_4)	0.021	0.420	-0.399	-0.740 (x_{11})
5	Cultivated Land Holding per capita (x_5)	-0.125	-0.640	0.515	1.027 (x_{11})
6	No. of land fragment (x_6)	-0.036	0.754	-0.790	0.591 (x_{11})
7	Avg. size of land fragment (x_7)	-0.118	0.803	-0.921	0.541 (x_{11})
8	Distance between land fragment (x_8)	0.204	0.104	0.100	-0.115 (x_7)
9	Homestead land per capita (x_9)	-0.106	-0.184	0.078	0.479 (x_{11})
10	Marketed Surplus per capita (x_{10})	0.086	0.149	-0.063	0.778 (x_{11})
11	On farm income per capita (x_{11})	-0.104	1.473	-1.577	-1.133 (x_{12})
12	Family Expenditure per capita (x_{12})	-0.191	-1.277	1.086	1.307 (x_{11})
13	Savings per capita (x_{13})	0.023	-0.805	0.828	1.082 (x_{11})
14	Land under irrigation per capita (x_{14})	-0.096	0.473	-0.569	1.011 (x_{11})
15	Cropping intensity (x_{15})	0.046	-0.076	0.122	0.454 (x_{11})
16	Livestock count (x_{16})	-0.305	-1.260	0.955	0.570 (x_{11})
17	Communication variable (x_{17})	-0.379	-0.232	-0.147	-0.248 (x_6)
18	LPG Consumption per person (x_{18})	-0.040	0.208	-0.248	0.698 (x_{11})
19	Production of organic manure (x_{19})	-0.111	0.204	-0.315	-0.995 (x_{16})
20	Application of organic manure (x_{20})	-0.130	-0.238	0.108	-0.992 (x_{16})
21	Rice stubble height (x_{21})	-0.018	0.086	-0.104	-0.811 (x_{16})
22	Crop residue left in field (x_{22})	-0.063	0.533	-0.596	1.005 (x_{11})
23	Soil conservation perception (x_{23})	-0.266	-0.036	-0.230	0.145 (x_7)
24	Water conservation perception (x_{24})	-0.193	0.011	-0.204	-0.181 (x_{12})
25	Bio-diversity conservation perception (x_{25})	-0.144	-0.023	-0.121	-0.462 (x_{12})
26	Appl. of climate resilient technology (x_{26})	-0.261	0.139	-0.400	-0.294 (x_{16})

Residual effect: *0.5182*

Highest individual indirect effect: ***On farm income per capita (x_{11})*** **[Frequency: 16]**

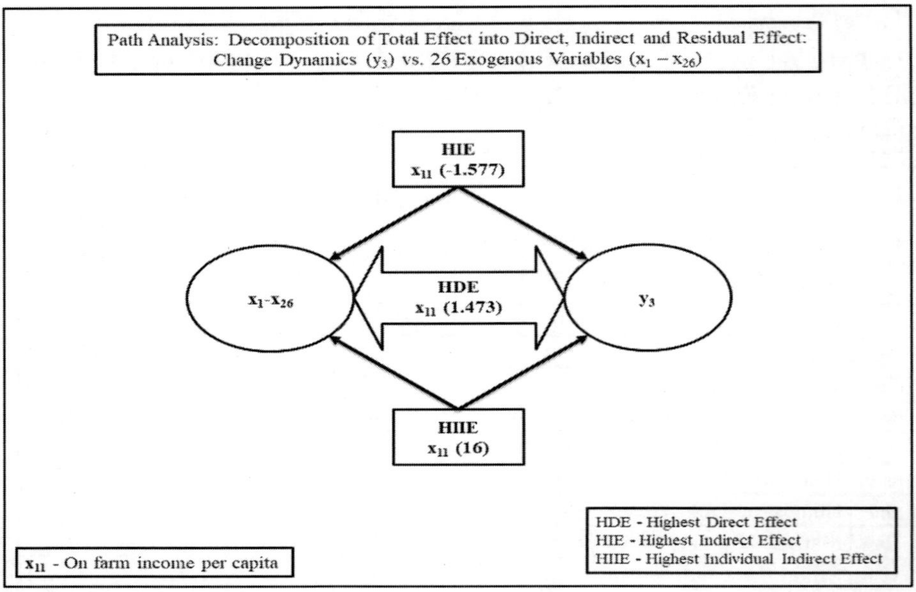

Model 9: Path Analysis: y_3 vs. $x_1 - x_{26}$

Result: Table 25presents the path analysis which has decomposed the overall impact of exogenous variables on the consequent variable into direct, indirect, and residual effects.

Revelation: The exogenous variable On farm income per capita (x_{11}) has recorded the highest direct effect on the consequent variable Change Dynamics (y_3). It certainty concludes that on farm income has been the strongest predictor for estimating the trend of agricultural change.

The variable has also exerted the highest indirect effect on the consequent variable by involving as many as 25 other exogenous variables. So, this variable has got significant strategic importance.

The exogenous variable, Communication variable (x_{17}) has been found to exert the highest total effect on the consequent variable Change Dynamics (y_3).

The residual effect being 51.8 percent, it is to conclude that with a combination of 26 exogenous variables, 51.8 percent of variance in the consequent variable, Change Dynamics (y_3) could not be explained.

6.14 Canonical Covariate Analysis: Interaction between Left Side Dependent and Right Side Independent Variables

Table 26: Canonical Cross Loading values of Dependent variables

Dependent Variable	**Coefficients**
Perception on Climate Change (y_1)	-0.03932
Perceived Uncertainty (y_2)	-0.15391
Change Dynamics (y_3)	0.24269

Table 27: Canonical Cross Loading values of Independent variables

Independent Variable	**Coefficients**
Age (x_1)	0.23511
Education (x_2)	1.30037
Functional education (x_3)	-0.69736
Size of family (x_4)	2.60925
Cultivated Land Holding per capita (x_5)	5.75125
No. of land fragment (x_6)	2.11499
Avg. size of land fragment (x_7)	1.74579
Distance between land fragment (x_8)	-0.06168
Homestead land per capita (x_9)	0.20800
Marketed Surplus per capita (x_{10})	-0.32865
On farm income per capita (x_{11})	0.67998
Family Expenditure per capita (x_{12})	-0.32863
Savings per capita (x_{13})	-0.79543
Land under irrigation per capita (x_{14})	-5.40577
Cropping intensity (x_{15})	1.24052
Livestock count (x_{16})	-2.74749
Communication variable (x_{17})	0.58299
LPG Consumption per person (x_{18})	0.23813
Production of organic manure (x_{19})	-5.19991
Application of organic manure (x_{20})	5.04458
Rice stubble height (x_{21})	0.07881
Crop residue left in field (x_{22})	0.49148
Soil conservation perception (x_{23})	0.33659
Water conservation perception (x_{24})	0.25844
Bio-diversity conservation perception (x_{25})	0.48476
Appl. of climate resilient technology (x_{26})	0.56786

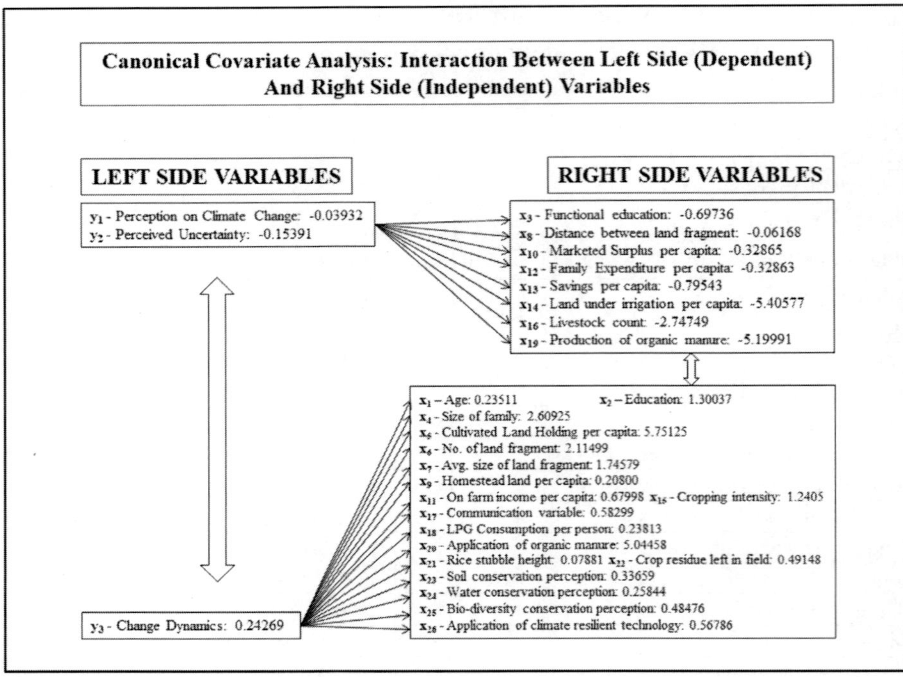

Model 10: CCA: Interaction between Left Side Dependent and Right Side Independent Variables

Result: Canonical covariant analysis is a valuable tool for comprehending and visually representing the relationship between subgroups of x and y variables. This analysis highlights the clear and distinct co-integrations of these variables, making them readily apparent and easily observable.

The Table 26 and 27 presents the results of the Canonical Covariate Analysis (CCA) where we have two sets of variable in parallel interaction based on cross loadings.

Revelation: It has been found out that two left side variables (dependent variables) y_1, $y_2$are moving together to form a homogenous sub-conglomeration of y variables. Empirically, the perception on climate change and perceived uncertainty do form a close kinship. Uncertainty is basically a reflection of climate change and a perception generates therewith. It shows that the diode of y_1, y_2 which are being responded to by the following variables - Functional education (x_3), Distance between land fragment (x_8), Marketed Surplus per capita (x_{10}), Family Expenditure per capita (x_{12}), Savings per capita (x_{13}), Land under irrigation per capita (x_{14}), Livestock count (x_{16}), Production of organic manure (x_{19}).

Similarly, change dynamics in a solitary way, when moves alone, it receives responses and impact from the following right side independent variables - Age (x_1),

Education (x_2), Size of family (x_4), Cultivated Land Holding per capita (x_5), No. of land fragment (x_6), Avg. size of land fragment (x_7), Homestead land per capita (x_9), On farm income per capita (x_{11}), Cropping intensity (x_{15}), Communication variable (x_{17}), LPG Consumption per person (x_{18}), Application of organic manure (x_{20}), Rice stubble height (x_{21}), Crop residue left in field (x_{22}), Soil conservation perception (x_{23}), Water conservation perception (x_{24}), Bio-diversity conservation perception (x_{25}), Appl. of climate resilient technology (x_{26}).

6.15 Factor Analysis: Strategic Conglomeration of Variables into Factor

Table 28: Extraction of factors through Principal Component Analysis

Factors	Variables	Factor loading	% of variance	Cumulative %	Factor renamed
F-1	Age (x1)	-0.572	30.193	30.193	Family Ecosystem
	Functional education (x3)	0.548			
	Size of family (x4)	-0.660			
	Cultivated Land Holding per capita (x5)	0.962			
	Avg. size of land fragment (x7)	0.574			
	Marketed Surplus per capita (x10)	0.672			
	On farm income per capita (x11)	0.790			
	Family Expenditure per capita (x12)	0.649			
	Savings per capita (x13)	0.706			
	Land under irrigation per capita (x14)	0.942			
	Livestock count (x16)	0.617			
	Crop residue left in field (x22)	0.912			
F-2	No. of land fragment (x6)	0.633	16.783	46.976	Ecological Resilience
	Communication variable (x17)	-0.466			
	LPG Consumption per person (x18)	-0.632			
	Production of organic manure (x19)	0.744			
	Application of organic manure (x20)	0.720			
	Rice stubble height (x21)	0.603			
	Perceived Uncertainty (y2)	-0.625			

Factors	Variables	Factor loading	% of variance	Cumulative %	Factor renamed
F-3	Education (x2)	-0.754	10.438	57.414	Conservation Dynamics
	Soil conservation perception (x23)	0.547			
	Water conservation perception (x24)	0.405			
	Appl. of climate resilient technology (x26)	0.646			
F-4	Homestead land per capita (x9)	-0.608	5.461	62.874	Eco-diversity Perception
	Bio-diversity conservation perception (x25)	0.443			
F-5	Distance between land fragment (x8)	0.530	4.739	67.614	NA
F-6	Perception on Climate Change (y1)	0.511	4.199	71.813	Climate Dynamics
	Change Dynamics (y3)	0.602			
F-7	Cropping intensity (x15)	-0.630	3.682	75.494	NA

Result: Principal Component Analysis (PCA) is the foundation of Factor Analysis, a technique used to consolidate operationally similar variables into strategic components. PCA serves as a dimension reduction approach, grouping a set of seemingly distinct variables into a unified factor. The efficacy of this consolidation is determined by the factor loading value, with higher values indicating stronger correlations among variables.

By applying Factor Analysis, the total number of interacting variables has been organized into identifiable factors based on factor loading and Eigen values. Each factor comprises of homogeneous variables that are closely interconnected, facilitating informed decision-making for effective resource and organizational management.

Revelation: From the Table 28, it has been found that Factor-1 has accumulated the following variables - Age (x_1), Functional education (x_3), Size of family (x_4), Cultivated Land Holding per capita (x_5), Avg. size of land fragment (x_7), Marketed Surplus per capita (x_{10}), On farm income per capita (x_{11}), Family Expenditure per capita (x_{12}), Savings per capita (x_{13}), Land under irrigation per capita (x_{14}), Livestock count (x_{16}) and Crop residue left in field (x_{22}); it has contributed 30.193% of variance into the total interaction. This factor has been renamed as Family Ecosystem.

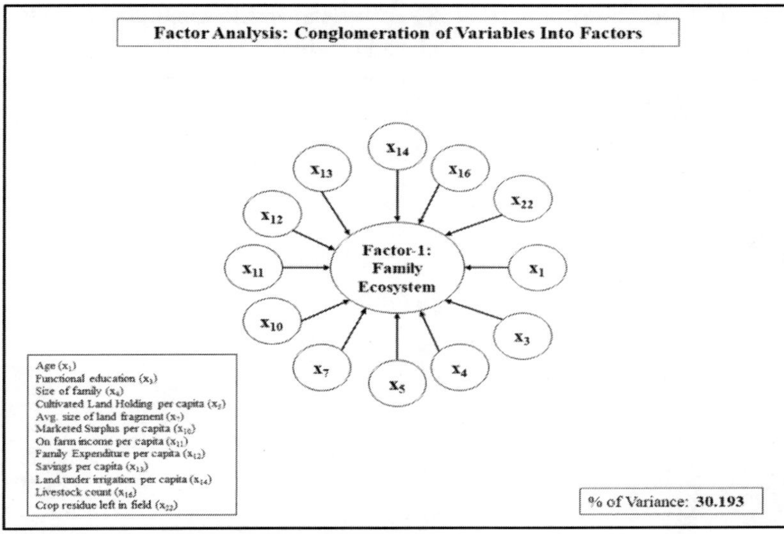

Model 11: Factor-1: Family Ecosystem

Similarly, Factor-2 has accumulated the following variables -No. of land fragment (x_6), Communication variable (x_{17}), LPG Consumption per person (x_{18}), Production of organic manure (x_{19}), Application of organic manure (x_{20}), Rice stubble height (x_{21}) and Perceived Uncertainty (y_2); it has contributed 16.783% of variance into the total interaction. This factor has been renamed as Ecological Resilience.

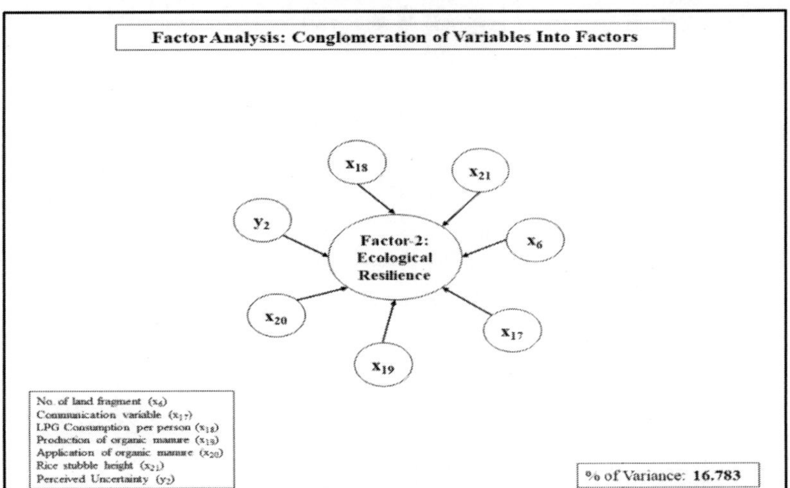

Model 12: Factor-2: Ecological Resilience

Factor-3 has accumulated the following variables -Education (x_2), Soil conservation perception (x_{23}), Water conservation perception (x_{24}) and Appl. of climate resilient technology (x_{26}); it has contributed 10.438% of variance into the total interaction. This factor has been renamed as Conservation Dynamics.

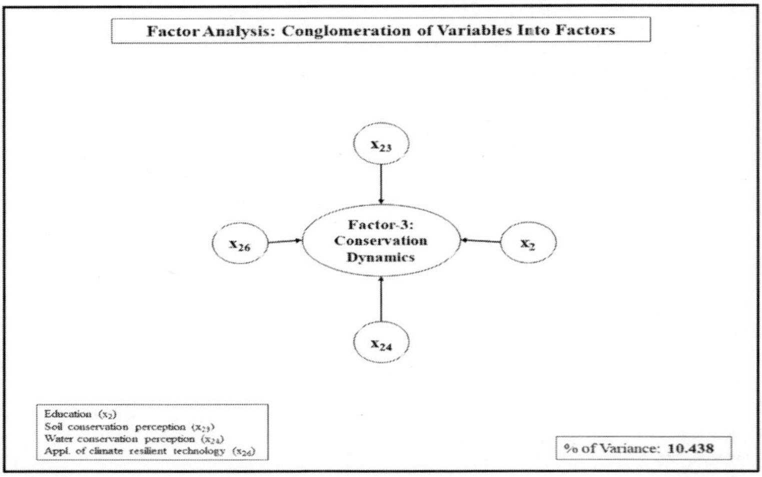

Model 13: Factor-3: Conservation Dynamics

Factor-4 has accumulated the following variables -Homestead land per capita (x_9) and Bio-diversity conservation perception (x_{25}); it has contributed 5.461% of variance into the total interaction. This factor has been renamed as Eco-diversity Perception.

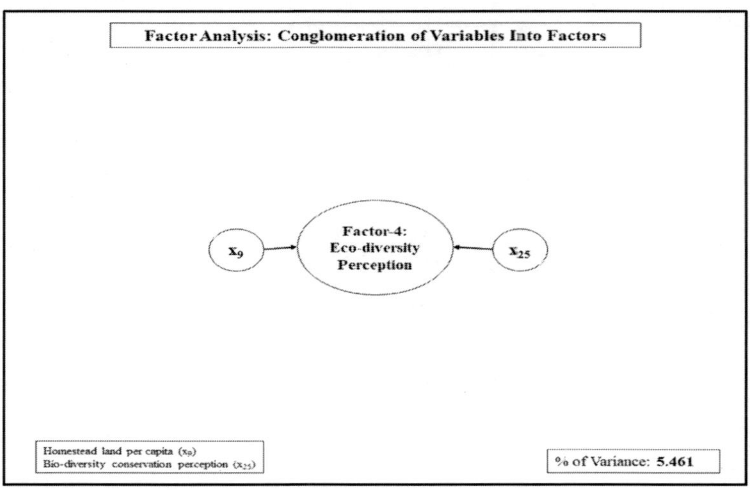

Model 14: Factor-4: Eco-diversity Perception

Factor-5 is consist of only one variable that is Distance between land fragment (x_8) and renaming is not necessary; it has contributed 4.739% of variance into the total interaction.

Factor-6 has accumulated the following variables -Perception on Climate Change (y_1) and Change Dynamics (y_3); it has contributed 4.199% of variance into the total interaction. This factor has been renamed as Climate Dynamics.

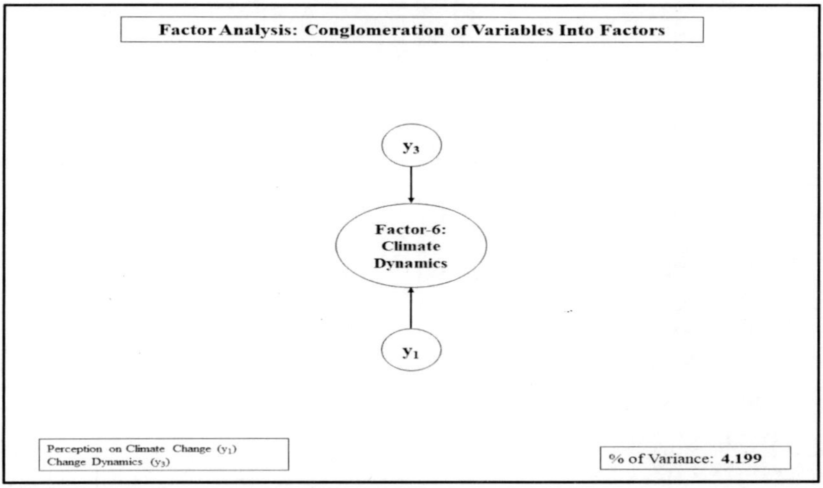

Model 15: Factor-6: Climate Dynamics

Factor-7 is consist of only one variable that is Cropping intensity (x_{15})and renaming is not necessary; it has contributed 3.682% of variance into the total interaction.

The Factor Analysis thus helps managing higher number of variables by extracting into deal able number of factors. All these factors has strategic importance.

6.16 Artificial Neural Network (ANN)

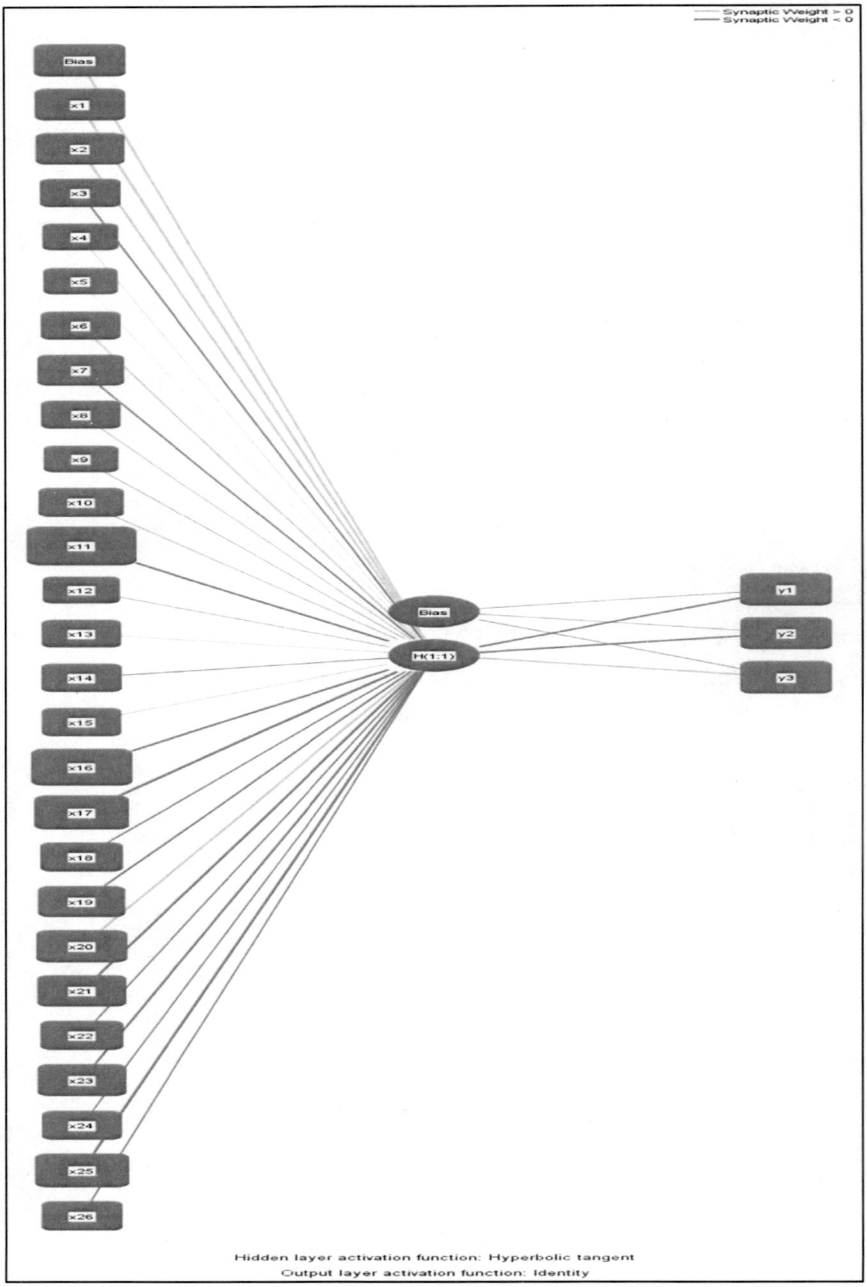

Model 16: Artificial Neural Network (Multilayer Perceptron)

Table 29: Model Summary of ANN

Model Summary			
Training	Sum of Squares Error		77.455
	Average Overall Relative Error		0.807
	Relative Error for Scale Dependents	Perception on Climate Change	0.799
		Perceived Uncertainty	0.756
		Change Dynamics	0.865
Testing	Sum of Squares Error		24.961
	Average Overall Relative Error		0.816
	Relative Error for Scale Dependents	Perception on Climate Change	0.826
		Perceived Uncertainty	0.856
		Change Dynamics	0.775

Table 30: Parameter Estimates of ANN

Parameter Estimates					
Predictor		**Predicted**			
		Hidden Layer 1	**Output Layer**		
		H(1:1)	y_1	y_2	y_3
Input Layer	(Bias)	.340			
	x_1	.415			
	x_2	.362			
	x_3	-.180			
	x_4	.046			
	x_5	.003			
	x_6	.268			
	x_7	-.236			
	x_8	.161			
	x_9	-.012			
	x_{10}	.275			
	x_{11}	-.470			
	x_{12}	-.039			
	x_{13}	.085			
	x_{14}	-.106			
	x_{15}	.105			
	x_{16}	-.481			
	x_{17}	-.688			
	x_{18}	-.211			
	x_{19}	-.291			

Parameter Estimates					
Predictor		Predicted			
		Hidden Layer 1	Output Layer		
		H(1:1)	y_1	y_2	y_3
	x_{20}	.455			
	x_{21}	-.389			
	x_{22}	-.151			
	x_{23}	-.336			
	x_{24}	-.140			
	x_{25}	-.382			
	x_{26}	-.145			
Hidden Layer 1	(Bias)		-.080	-.067	-.089
	H(1:1)		-.528	-.572	.679

Table 31: Independent Variable Importance

Independent Variable Importance		
	Importance	**Normalized Importance**
Age (x_1)	.058	46.3%
Education (x_2)	.052	42.2%
Functional education (x_3)	.024	19.5%
Size of family (x_4)	.006	5.2%
Cultivated Land Holding per capita (x_5)	.001	0.5%
No. of land fragment (x_6)	.019	15.6%
Avg. size of land fragment (x_7)	.045	35.9%
Distance between land fragment (x_8)	.019	15.1%
Homestead land per capita (x_9)	.002	1.8%
Marketed Surplus per capita (x_{10})	.038	30.5%
On farm income per capita (x_{11})	.124	100.0%
Family Expenditure per capita (x_{12})	.010	8.4%
Savings per capita (x_{13})	.018	14.8%
Land under irrigation per capita (x_{14})	.022	17.5%
Cropping intensity (x_{15})	.018	14.8%
Livestock count (x_{16})	.095	76.4%
Communication variable (x_{17})	.071	57.6%
LPG Consumption per person (x_{18})	.029	23.3%
Production of organic manure (x_{19})	.045	36.2%
Application of organic manure (x_{20})	.058	46.6%

Independent Variable Importance		
	Importance	Normalized Importance
Rice stubble height (x_{21})	.051	41.2%
Crop residue left in field (x_{22})	.030	24.5%
Soil conservation perception (x_{23})	.051	40.9%
Water conservation perception (x_{24})	.020	16.4%
Bio-diversity conservation perception (x_{25})	.069	55.3%
Appl. of climate resilient technology (x_{26})	.023	18.7%

Result

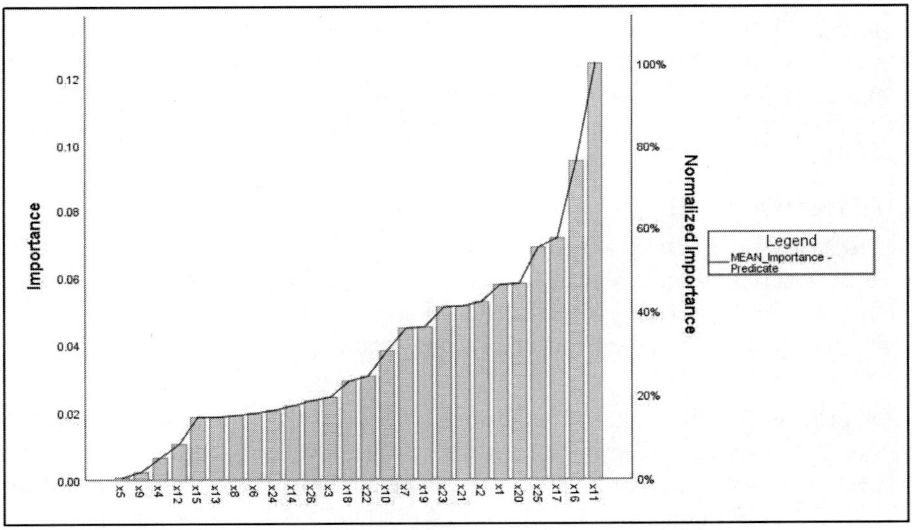

Graph 1: Presentation of Independent Variable Importance through Bar graph

Revelation: Model 16 presents Artificial Neural Network (ANN) diagram.

It is basically mimicry of human brain and neuron in operation. Artificial Neural Network has been carried out in the 'IBM SPSS Statistics version 25' software to measure the efficacy of input variables (x_1-x_{26}), passing through the hidden layers and ultimately impacting on the output variables (y_1-y_3). ANN has got 3 layers viz. input layer, hidden layer and output layer. Through the hidden layers, errors are corrected. The number of hidden layer depends on the amount of errors as estimated for the training of input data. Hidden layers are made functional by adding bias to it to minimize the error. The input variables passing through the hidden layers and impacting onto the output variables dominantly have been figured out and indicated by the bold coloured lines.

Graph 1 shows the normalized importance of the input variables and considering the normalized importance, it has been found out that the variable On farm income (x_{11}) has exaggerated the highest functional importance on output variables. So, climate mitigation and adaptation through a changed function and attitude needs to redress On farm income related issues. Farmers must be put into the understanding of climate adaptation and mitigation strategies. So that, they can secure better on farm income and higher dividend.

It has also been found that the input variable Livestock count (x_{16}) has got dominant impact on output variables followed by Communication variable (x_{17}), Bio-diversity conservation perception (x_{25}), Application of organic manure (x_{20}), Age (x_1), Education (x_2), Rice stubble height (x_{21}) and so on. Both Livestock count (x_{16}) and Application of organic manure (x_{20}) have taken the spear heading role in the ambit of climate change and perception. When ongoing farms are persistently receiving organic manure, it becomes easier to deal with the brunt of climate change. Livestock is the provider of regular family income also. Thus it helps to get over the climate change effect brunt with higher economic capabilities and resilience.

6.17 Mental Modelling – Fuzzy Logic Cognitive Mapping

The present study has tried to visualize the farmers' perception on climate change, level of uncertainty and trend of agricultural change in the selected study area using "Mental Modeler" software. Mental models on farmers' perception on different aspects were created through Fuzzy-logic Cognitive Mapping (FCM) for the research locale.

After processing, cleaning and analysing all the collected data, most relevant factors related to the perception on climate change, level of uncertainty and trend of agricultural change according to the respondents were selected to visualize their cognitive state of mind through FCM with the help of 'Mental Modeler' software.

Results: A shared mental map was generated by conglomerating each respondents' fuzzy cognitive maps from which the issue of the research interest in the studied villages can be understood.

The node property and network property of the shared mental map through network analysis were examined from which the driver, ordinary and receiver components were isolated by studying the in-degree, out-degree and centrality values of each component.

6.17.1 Shared Mental Map of multiple respondents

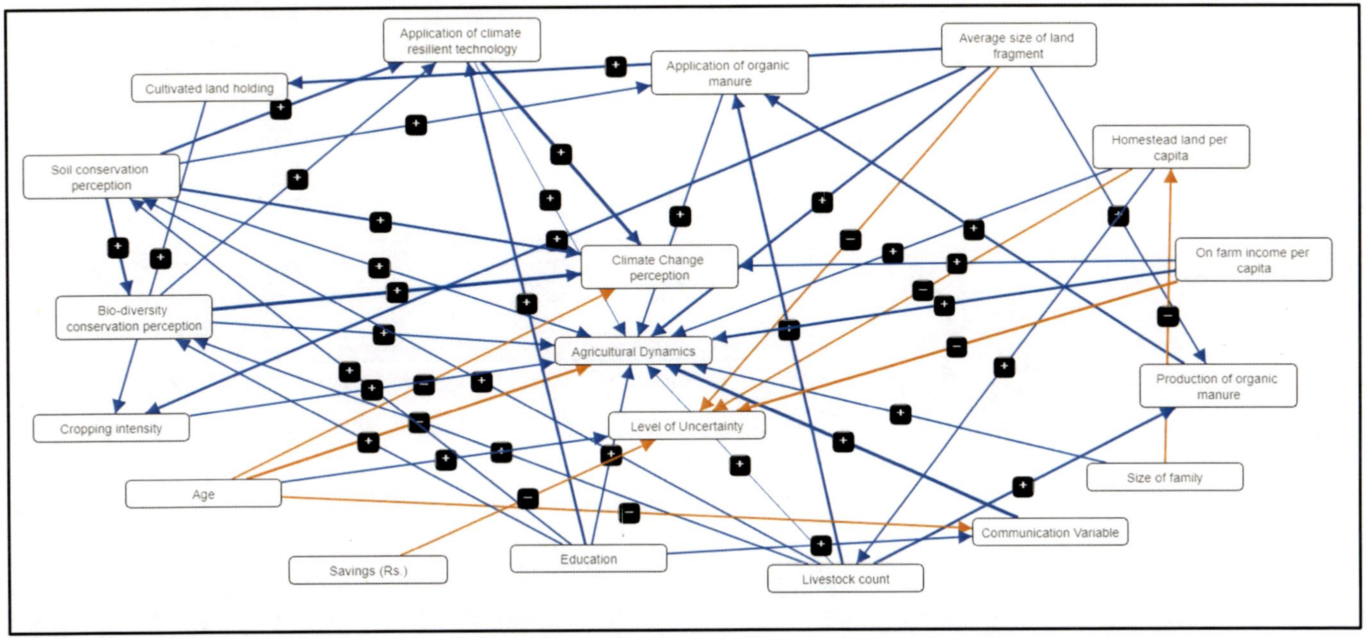

Model 17: Mental Model of multiple respondents on agricultural dynamics, climate change perception and level of uncertainty

This one map combines all the separate maps made based on the responses of multiple respondents. A clear picture can be obtained on the topic of research interest based on all the possible elements which are present in the map where there are 19 elements and 43 connections between them.

6.17.2 Network Analysis

Table 32: Network Properties

Total Components	Total Connections	Density	Connections per Component	No. of Driver Components	No. of Receiver Components	No. of Ordinary Components	Complexity Score
19	43	0.13	2.26	6	3	10	0.5

Table 33: Node Properties

Component	In-degree	Out-degree	Centrality	Type
Average size of land fragment	0.00	2.60	2.60	driver
Homestead land per capita	0.20	0.86	1.06	ordinary
Application of climate resilient technology	1.70	1.01	2.71	ordinary
Bio-diversity conservation perception	1.05	1.74	2.79	ordinary
Communication Variable	0.50	1.00	1.50	ordinary
Production of organic manure	0.80	0.50	1.30	ordinary
Soil conservation perception	0.50	2.71	3.21	ordinary
Size of family	0.00	0.48	0.48	driver
Livestock count	0.30	1.68	1.98	ordinary
Application of organic manure	1.30	0.23	1.53	ordinary
Age	0.00	1.47	1.47	driver
Agricultural Dynamics	5.39	0.00	5.39	receiver
On farm income per capita	0.00	1.90	1.90	driver
Climate Change perception	3.40	0.00	3.40	receiver
Level of Uncertainty	1.80	0.00	1.80	receiver
Education	0.00	1.58	1.58	driver
Cropping intensity	1.10	0.33	1.43	ordinary
Cultivated land holding	0.70	0.45	1.15	ordinary
Savings (Rs.)	0.00	0.20	0.20	driver

Driver components: The Model 17 and Table 33 depicts that components like Average size of land fragment, Age, On farm income per capita and Education have significant out-degree value with zero (0) in-degree value, thus making them system drivers. So these driver components contribute significantly on perception on climate change, level of uncertainty and trend of agricultural change.

The Table 33 also shows Size of family and Savings (Rs.) as drivers, however these components should not be considered as system drivers because of low out-degree value. So these components are the other reasons that contribute in the whole process.

Ordinary Components: It is well discernible that components like Application of climate resilient technology, Application of organic manure and Cropping intensity have remarkable in-degree value with certain out-degree value.

Application of climate resilient technology has the highest in-degree value and a high centrality value among ordinary components because it is connected to many other components. It is one of the majorcomponent that affects farmers' perception on climate change.

Another crucial component, Bio-diversity conservation perception has significant in-degree and out-degree value resulting in high centrality value. While components like Soil conservation perception, Livestock count have significant out-degree value with certain in-degree value that also results in high centrality value.

Receiver components: It can be inferred from the Table 33 that Agricultural Dynamics, Climate Change perception and Level of Uncertainty are the receiver component of the system and all the components have significant in-degree value with zero (0) out-degree value and the receiver component Agricultural Dynamics has the highest centrality value as it is connected with many of the ordinary and driver components.

Revelation: Climate change perception is associated with uncertainty and uncertainty characterises the agricultural dynamics. The shared mental map displays that the episodes of climate change perception and uncertainty are mainly generating from Average size of land fragment, Age, On farm income per capita and Education which are considered as the driver in the 'Mental Modeler' suite.

Farmers are more concerned about their income in the scenario of rapid changing climate which has already enhanced disruptions in income. Income again is related to the size of land holding and perception correlates with education. While accepting a micro-sociological policy, the young farmers with classified land fragment and income category play a pivotal role.

It is noticeable from field level exposure that farmers of young age mostly avoid field operations when temperature turns extreme. The fragmented holdings drive the farmers to become cost and energy prodigal.

Thus the shared mental map as well as the analysed network and node properties nicely depicts the shared perception on impact of climate change on income which ultimately elicits the change in agricultural dynamics decisively.

6.17.3 Scenario Analysis

Scenario Analysis was conducted as shown in Graph 2 and Graph 3 using the adjacency matrix. It was performed in order to graphically visualize the detailed scenario using Mental Modeler's Scenario Analysis Module, taking into account the viewpoint of the respondents as well as their current situation. In both the cases, hyperbolic tangent transfer function was deployed.

Scenario-1: Preferred State

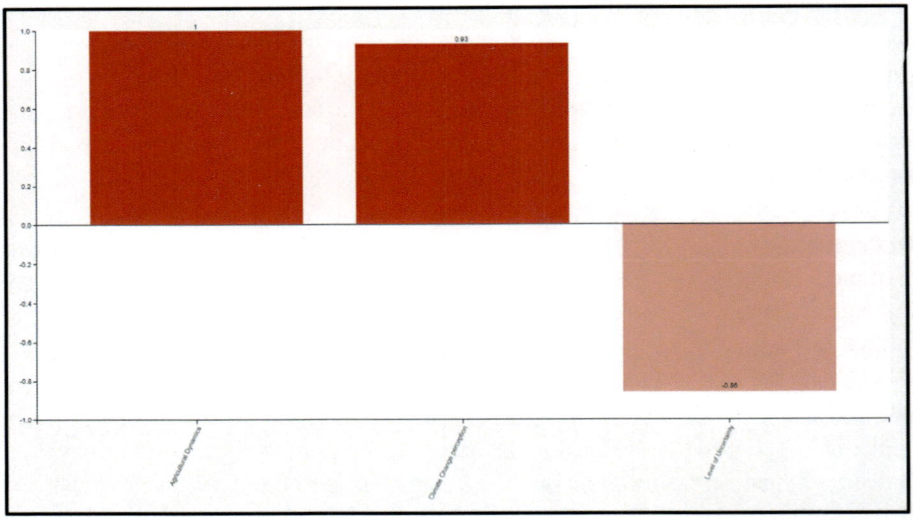

Graph 2: Preferred State scenario

Scenario-2: Actual State

Graph 3: Actual State scenario

Results

Scenario-1: In case of scenario-1, Agricultural Dynamics has shown an increase while Level of Uncertainty has shown a decrease which is preferable in general by multiple respondents as portrayed in the shared mental map and in between these two receiver component, the Climate Change Perception has also shown an increase.

Scenario-2: In case of scenario-2 which depicts the actual state of the multiple respondents, it is noticeable that everything does not go as preferred in the real world situation irrespective of perception, knowledge, desire etc. It has been found that both the receiver component, Agricultural Dynamics and Climate Change Perception has shown a decrease and interestingly the Level of Uncertainty has also shown a decrease.

Revelation: The scenario-1 presents the graphical state of preferred choices of the respondents in order to deal with perceived climate change variability. It shows that agricultural dynamics have a positive (1) and climate change perception also have a positive (0.93) change whereas level of uncertainty goes down by -0.86 units. The micro-sociological policy advocates that in case we want to ameliorate uncertainties from the farmers' mind-set, we need to build up proper climate change perception amongst farmers so to set the agriculture into the dynamics of adaptive and resilient change.

In another scenario (scenario-2), the heuristics move in a unique way wherein climate change perceptions goes down (-0.56), agricultural dynamics goes down (-0.63) and uncertainty also interestingly moves down negatively (-0.31). Perhaps this is the reality of farmers with low or no concern about climate change. Thus the level of uncertainty moves in a negative direction. This kind of scenario is mostly depicting that the farmers are losing interest in farming as a whole or are preferring to go stoic in a situation whatever may the outcome. This is not the ideal condition. It seems respondents are fatalistic, believing in fortune and fate. To protect our agriculture in the face of climate change reality, we have to reduce their perceived uncertainty by making agricultural dynamics more adaptive, regenerative, remunerative and productive to ameliorate the brunt of climate change.

7

Epilogue

7.1 Summary

Climate change is a complex global phenomenon that poses significant challenges for agriculture including uncertainty about future weather patterns, soil conditions, and other environmental factors. Farmers are facing the difficult task of interpreting and responding to these changes in order to maintain their livelihoods and ensure food security for growing populations. In this context, mental modelling has emerged as a useful approach for understanding how farmers interpret and respond to climate change. Mental models are cognitive representations of how individuals understand and make decisions about complex systems, and can be used to identify key drivers of behaviour and decision-making.

The present study has explored the social, ecological, and economic factors that influence farmers' mental models on perception of climate change, level of uncertainty and trend of agricultural change in the selected study area. It argues that farmers' mental models are shaped by a range of factors, including their own experiences and observations, social networks and interactions, access to information and resources, and broader economic and political contexts. The present study has also highlighted the role of uncertainty in shaping farmers' mental models and decision-making. Uncertainty about the future impacts of climate change can create significant challenges for farmers, as they must make decisions based on incomplete information and uncertain predictions.

The current research work was carried out in the new alluvial zone which is one of the agro-ecological regions in the state of West Bengal. To conduct the study, three villages located in the Haringhata block of Nadia district were chosen randomly. A total of 90 farmers were selected with 30 respondents each from the villages of Kastadanga, JalkarBhomra and Subuddipur. Initially a list of contact farmers was collected from the Kastadanga SKUS and few contact farmers were selected randomly from the list. Afterwards Snowball sampling method was followed to select respondents.

The primary data was collected using a structured interview schedule through personal interviews. In addition, secondary data was gathered from various sources such as departmental library and Central Library, Bidhan Chandra Krishi Viswavidyalaya; internet and the Department of Agriculture, GoWB and GoIin order to establish the conceptual framework of the study.

The general objective of the present research is to study the "Climate Change, Uncertainty and Agricultural Dynamics: Social, Ecological and Economical Interpretation by Farmer through Mental Modelling Approach."

The specific research objectives of the present study are

i. To estimate the farmers' perception on climate change, level of uncertainty and trend of agricultural change for the selected study area.
ii. To identify and customize the pertinent variables present in the selected study area.
iii. To elucidate and evaluate the intra and inter level interactions among and between the sets of exogenous variables and consequence variables.
iv. To formulate a fuzzy logic cognitive map by using mental modelling approach to visualize farmers' perception on climate change, uncertainty and agricultural dynamics.
v. To generate micro-sociological policy implications based on the empirical research.

For the study, a total of 26 independent variables and 3 dependent variables were chosen. These variables were measured using precise scales that were either developed by previous researchers in the field of social science or modified versions of existing scales, tailored specifically for the requirements of this study. The measurements were obtained through a structured interview schedule.

The independent variables selected for the study are as follows - Age (x_1), Education (x_2), Functional education (x_3), Size of family (x_4), Cultivated Land Holding per capita (x_5), No. of land fragment (x_6), Avg. size of land fragment (x_7), Distance between land fragment (x_8), Homestead land per capita (x_9), Marketed Surplus per capita (x_{10}), On farm income per capita (x_{11}), Family Expenditure per capita (x_{12}), Savings per capita (x_{13}), Land under irrigation per capita (x_{14}), Cropping intensity (x_{15}), Livestock count (x_{16}), Communication variable (x_{17}), LPG Consumption per person (x_{18}), Production of organic manure (x_{19}), Application of organic manure (x_{20}), Rice stubble height (x_{21}), Crop residue left in field (x_{22}), Soil conservation perception (x_{23}), Water conservation perception (x_{24}), Bio-diversity conservation perception (x_{25}), Appl. of climate resilient technology (x_{26}).

The dependent variables selected for the study are as follows - Perception on Climate Change (y_1), Perceived Uncertainty (y_2) and Change Dynamics (y_3).

Findings

(I) The dependent variables were used to estimate the farmers' perception on climate change, level of uncertainty and trend of agricultural change for the selected study area. The dependent variables were measured using Likert's Scale that was already in existence as well as by using modified versions of that existing scale, tailored specifically for the requirements of this study. The measurements were obtained through a structured interview schedule followed by indexing through scoring.

The estimated scores of the dependent variables were undergone Z standardization before data analysis to have a mean of 0 and a standard deviation of 1. So that the whole dataset can be compared on a common scale and so to employ the dataset in statistical analysis and machine learning algorithms.

(II) Statistical tools such as Coefficient of correlation, Stepwise Regression Analysis, Path Analysis and Factor Analysis (Principal Component Analysis) were deployed in the present research work. Based on outcomes of the analysis, pertinent variables operating in the research locale were identified by 'r' and R^2 values through Coefficient of correlation and Stepwise Regression Analysis respectively; customised based on the direct, indirect effect through Path analysis as well as factor renaming through Factor Analysis.

Coefficient of Correlation (r): Perception on Climate Change (y_1) vs. 26 Independent Variables (x_1 – x_{26})depicts that Age (x_1), Size of family (x_4), Cultivated Land Holding per capita (x_5), Avg. size of land fragment (x_7), Marketed Surplus per capita (x_{10}), On farm income per capita (x_{11}), Family Expenditure per capita (x_{12}), Land under irrigation per capita (x_{14}), LPG Consumption per person (x_{18}), Crop residue left in field (x_{22}), Soil conservation perception (x_{23}), Bio-diversity conservation perception (x_{25}) and Appl. of climate resilient technology (x_{26})have recorded significant correlation with the dependent variable Perception on Climate Change (y_1).

Coefficient of Correlation (r): Perceived Uncertainty (y_2) vs. 26 Independent Variables (x_1 – x_{26})presents that Age (x_1), Functional education (x_3), Size of family (x_4), Avg. size of land fragment (x_7), Homestead land per capita (x_9), On farm income per capita (x_{11}), Family Expenditure per capita (x_{12}), Communication variable (x_{17}), LPG Consumption per person (x_{18}), Production of organic manure (x_{19}), Application of organic manure (x_{20}), Soil conservation perception (x_{23}), Water conservation perception (x_{24}), Bio-diversity conservation perception (x_{25}), Appl. of climate resilient technology (x_{26}) have recorded significant correlation with the dependent variable Perceived Uncertainty (y_2).

Coefficient of Correlation (r): Change Dynamics (y_3) vs. 26 Independent Variables (x_1 – x_{26})shows that Livestock count (x_{16}), Communication variable (x_{17}), Soil conservation perception (x_{23}), Appl. of climate resilient technology (x_{26}) have recorded significant correlation with the dependent variable Change Dynamics (y_3).

Stepwise Regression Analysis: Perception on Climate Change (y_1) vs. 26 Causal Variables (x_1 – x_{26}) depicts that Appl. of climate resilient technology (x_{26}), Bio-diversity conservation perception (x_{25}), Avg. size of land fragment (x_7) and Homestead land per capita (x_9) have been retained at the last step of the analysis and have substantially contributed to the consequent variable Perception on

Climate Change (y_1). The R square value being 33.7%, it can be inferred that these 4 variables together have contributed 71.09% $\left(\frac{33.7}{47.4}\times 100\right)$ to the totalvariance in $y_{1,}$ thus making them functionally important causal variables operating in the study area.

Stepwise Regression Analysis: Perceived Uncertainty (y_2) vs. 26 Causal Variables ($x_1 - x_{26}$) presents that Size of family (x_4), Soil conservation perception (x_{23}), Production of organic manure (x_{19}) and Bio-diversity conservation perception (x_{25}) have been retained in the last step of the analysis and have substantially contributed to the consequent variable Perceived Uncertainty (y_2). The R square value being 37.6%, it can be inferred that these 4 variables together have contributed 60.6% $\left(\frac{37.6}{62}\times 100\right)$ to the totalvariance in $y_{2,}$ thus making them functionally important causal variables operating in the study area.

Stepwise Regression Analysis: Change Dynamics (y_3) vs. 26 Causal Variables ($x_1 - x_{26}$) shows that Communication variable (x_{17}) and Livestock count (x_{16}) have been retained in the last step of the analysis and have substantially contributed to the consequent variable Change Dynamics (y_3). The R square value being 25.4%, it can be inferred that these 2 variables together have contributed 52.4% $\left(\frac{25.4}{48.4}\times 100\right)$ to the total variance in $y_{2,}$ thus making them functionally important causal variables operating in the study area.

Path Analysis: Decomposition of Total Effect into Direct, Indirect and Residual Effect: Perception on Climate Change (y_1) vs. 26 Exogenous Variables ($x_1 - x_{26}$) depicts that Appl. of climate resilient technology (x_{26}) has exerted the highest total effect and Cultivated Land Holding per capita (x_5) has exerted the highest direct effect as well as the highest indirect effect on the dependent variable Perception on Climate Change (y_1). Besides, Cultivated Land Holding per capita (x_5) has also recorded the highest individual indirect effect with a frequency of 21.

Path Analysis: Decomposition of Total Effect into Direct, Indirect and Residual Effect: Perceived Uncertainty (y_2) vs. 26 Exogenous Variables ($x_1 - x_{26}$) presents that Size of family (x_4) has exerted the highest total effect and Cultivated Land Holding per capita (x_5) has exerted the highest direct effect as well as the highest indirect effect on the dependent variable Perceived Uncertainty (y_2). Besides, Cultivated Land Holding per capita (x_5) has also recorded the highest individual indirect effect with a frequency of 22.

Path Analysis: Decomposition of Total Effect into Direct, Indirect and Residual Effect: Change Dynamics (y_3) vs. 26 Exogenous Variables ($x_1 - x_{26}$) shows that Communication variable (x_{17}) has exerted the highest total effect and On farm income per capita (x_{11}) has exerted the highest direct effect as well as the highest indirect effect on the dependent variableChange Dynamics (y_3). Besides,

On farm income per capita (x_{11}) has also recorded the highest individual indirect effect with a frequency of 16.

Factor Analysis: Strategic Conglomeration of Variables into Factor- The process of conglomerating operationally similar variables into strategic components has been done by Principal Component Analysis based on factor loading. Factor analysis, specifically Principal Component Analysis (PCA), is an approach used to reduce dimension or data. In this method, the total number of interacting variables is grouped into identifiable factors based on factor loading and Eigen values. Each factor comprises of homogeneous variables that are closely interconnected, enabling informed strategic decision-making for effective resource and organizational management.

All the 29 variables were extracted into 7 factors based on their factor loading through PCA. Factor-1, Factor-2, Factor-3, Factor-4, Factor-5, Factor-6 and Factor-7 has accommodated 12, 7, 4, 2, 1, 2, 1 variables respectively as well as has contributed 30.193%, 16.783%, 10.438%, 5.461%, 4.739%, 4.199%, 3.682% variance into the total interaction respectively. Factor-1, Factor-2, Factor-3, Factor-4 and Factor-6 has been renamed as Family Ecosystem, Ecological Resilience, Conservation Dynamics, Eco-diversity Perception and Climate dynamics while Factor-5 and Factor-7 are solitary factor, so there is no need of renaming.

(III) Statistical tools such as Canonical Covariate Analysis (CCA) as well as Machine Learning Algorithm such as Artificial Neural Network (ANN) were used in the data analysis of the present research work. The results of the data analysis shows the intra and inter level interactions among and between the sets of exogenous variables and consequence variables.

Canonical Covariate Analysis: Interaction between Left Side Dependent and Right Side Independent Variables depicts two sets of variable in parallel interaction based on cross loadings and highlights the clear and distinct co-integrations of these variables, making them readily apparent and easily observable.

Artificial Neural Network (ANN) has figured out the input variables passing through the hidden layers and impacting onto the output variables dominantly which have been indicated by the bold coloured lines as shown in the ANN model. Graph 1 shows the normalized importance of the input variables and considering the normalized importance, it has been found out that the variable On farm income (x_{11}) has exaggerated the highest functional importance on output variables followed by Livestock count (x_{16}), Communication variable (x_{17}), Bio-diversity conservation perception (x_{25}), Application of organic manure (x_{20}), Age (x_1), Education (x_2), Rice stubble height (x_{21}) and so on.

(IV) The present research work has deployed the mental modelling and fuzzy logic cognitive mapping frameworks to provide a deeper understanding of the complex systems pertinent in the research locale through the shared mental map of multiple respondents as well as through the network analysis.

It has been found through the shared mental map and the network analysis that Agricultural Dynamics, Climate Change perception and Level of Uncertainty having significant in-degree value are the receiver component as the research interest suggests. On the other hand, Average size of land fragment, Age, On farm income per capita and Education having significant out-degree value are the driver component of the system while components like Application of climate resilient technology, Application of organic manure and Cropping intensity have remarkable in-degree value with certain out-degree value, making them ordinary component.

The Scenario Analysis has graphically portrayed 2 different scenario and it has been found that in scenario-1 where preferred state of multiple respondents has been depicted, Agricultural Dynamics and Climate Change Perception have shown an increase while Level of Uncertainty has shown a decrease. On the other hand, it has been found in scenario-2 which depicts the actual state of the multiple respondents, Agricultural Dynamics and Climate Change Perception has shown a decrease and interestingly the Level of Uncertainty has also shown a decrease.

(V) The micro-sociological policy implications based on the present empirical study advocates that persistent failures of existing agriculture to generate happy returns and isochronous escalation of stressful uncertainty will make the farmers stoic than sensible to adaptive and mitigating interventions. They will turn more stoic to climate change and its impacts whatever may come. This is really a concern for ensuring food security by increasing yield making agriculture resilient.

7.2 Conclusion

Inter Governmental Panel on Climate Change (IPCC) in its 6th Annual Report expressed serious concern from the read direction and importance on agriculture, health and livelihoods of millions of people across the terrains of class, culture and community. The change dynamics are evolving and twisting very fast. The apprehended global warming has now been predicted to go up by 1.5° to 3.0°C at the turn of the century. Due to the global warming, productivity of all major crops (C3) will suffer except crops like millets which can survive better.

Over the centuries, there has been a long-standing conflict between sociology, ecology and economy. The advent of industrialization, followed by the modernization of agriculture, has brought about much-needed changes in human civilization. As agricultural productivity continues to rise, it is expected that there will be a sufficient market demand to absorb the surplus. With changing income dynamics, it is crucial to ensure that enough jobs are being created in the subsector. However, the reality is far harsher. India is currently facing an ecological catastrophe in the agricultural sector due to the excessive and ongoing exploitation of soil, water, and biodiversity. If not addressed, this situation could lead to severe disruptions in social life as well as the presence of strong push factors may lead to a significant migration of the rural population towards urban areas.

The present research has precisely identified several factors that contribute significantly either positively or negatively in sociology, ecology and economy of the research locale. The outcome of the present study based on the results found, can be concluded as follows -

(I) The present study has elucidated the responses of Perception on Climate Change (y_1), Perceived Uncertainty (y_2) and Change Dynamics (y_3) to a score of 26 independent variables so to estimate the farmers' perception on climate change, level of uncertainty and trend of agricultural change for the selected study area.

(II) The variables which have dominant influence and the variables which have tremendous functional impact on the farmers' perception on climate change, level of uncertainty and trend of agricultural change in the research locale are clearly noticeable through coefficient of correlation and stepwise regression analysis respectively such as Size of family (x_4), Avg. size of land fragment (x_7), Homestead land per capita (x_9), Livestock count (x_{16}), Communication variable (x_{17}), Production of organic manure (x_{19}), Soil conservation perception (x_{23}), Bio-diversity conservation perception (x_{25}), Appl. of climate resilient technology (x_{26}) etc.

Through a dimension reduction approach (PCA), the following factors have been extracted - Family Ecosystem, Ecological Resilience, Conservation Dynamics, Eco-diversity Perception and Climate dynamics. These factors being strategically important, can successfully be examined for delineating policy implications.

It is well discernible that through managing soil, water, bio-diversity, livestock, fragments of land and pro-people communication, we can ensure perhaps a better and wider deterrence to the brunt of climate change.

(III) The Canonical Covariate Analysis (CCA) has played a crucial role in illustrating the sub-grouping of the variables which may help developing a bilateral strategy/policy implications so to address issues related to the farmers' perception on climate change, level of uncertainty and trend of agricultural change at both micro and macro levels.

This empirical study has gone further by elucidating the factors, as evinced by Artificial Neural Network (ANN) which are the drivers of ecologically resilient interventions that may help to get over the brunt of climate change effect. However, better returns should be assured at the end else ecology without economy will remain illusory as a proposition.

Community based approach to agriculture can overcome limitation imposed by fragmentations. Sharing of natural resources including water and other common properties can help us to earn higher capabilities and resilience to deal with the impending threats of climate change.

(IV) To realize the full potential of Fuzzy Cognitive Mapping (FCM) frameworks, collaboration and knowledge sharing among researchers, practitioners, policymakers, and farmers are essential. Continued research and development efforts are needed to refine and validate the models, enhance user-friendliness, and adapt them to diverse agricultural systems and contexts.

The outcome of the scenario analysis may help enhancing decision-making processes so for policy implications for optimized resource allocation and promotion of sustainable practices in agriculture and its allied sectors.

(V) We have to take up greater roles to protect our operating agro-ecosystem by ensuring better returns from farms as well conserving soil, water and bio-diversity. Else, climate change and uncertainty generated therewith will swallow agricultural dynamics specifically the production and food security in a developing country like India with huge population.

8

Limitation of the Research

Despite the meticulous methodology and extensive analysis employed throughout this research project, it is crucial to acknowledge and address the limitations that arose during the study. By recognizing these limitations, we not only ensure a transparent evaluation of the research process but also shed light on potential areas for future investigations to refine and expand upon the findings.

In this section of the chapter, the limitations encountered in this thesis are highlighted as follows:

1. Due to the complexity of the concept, obtaining quantitative responses proved to be exceedingly challenging.
2. The development of improved scales for scoring could have enhanced the research if implemented.
3. The inclusion of a greater number of relevant variables could have provided a more comprehensive and in-depth understanding of the research topic.
4. Deploying mathematical optimization and simulations could have effectively integrated and co-integrated the seemingly disparate aspects of climate change perception, uncertainty perception, and perception of agricultural dynamics.
5. Presently, farmers are hesitant to engage in conversations and provide data to individuals outside their close social circles.

9

Future Scope of the Study

As the present study draws to a close, it opens up a myriad of possibilities for future research and exploration in the field of the present study. The insights gained and the findings presented in this thesis may provide a foundation for building upon and extending the existing body of knowledge. By identifying key areas that require additional attention and offering suggestions for future studies, this section of the chapter aims to inspire and guide researchers in their pursuit of enhancing our understanding and driving progress in this dynamic and evolving field.

The present study leaves behind the following domains which may be researched out in future:

1) The present research can be replicated with varied geo-spatial situations viz., in different locations and settings, with different respondents, with spatio-temporal expansion and methodological innovation.
2) PRA tools and materials can be deployed to collect more specific data to compensate for the inadequate information to go for the classical statistical approaches.
3) Multi-criteria decision analysis (MCDA) techniques can be integrated with mental modelling and fuzzy logic cognitive mapping so to enhance the decision-making process.
4) To realize the full potential of these frameworks, collaboration and knowledge sharing among researchers, practitioners, policymakers, and farmers are essential. Continued research and development efforts are needed to refine and validate the models, enhance user-friendliness, and adapt them to diverse agricultural systems and contexts.
5) A pool of big data distributed over 30 years or more can undergo simulation and decision ecology for promulgating micro sociological policy and application. For this, application of ARIMA (Autoregressive Integrated Moving average) and SEM (Structural Equation Modelling) as instruments of interpretation and analysis can be more effective.

Bibliography

AAAS. (2023). About Science and AAAS. Www.science.org. https://www.science.org/content/page/about-science-aaas

Admin. (2017, October 2). Deforestation. BYJUS; Byju's. https://byjus.com/chemistry/deforestation/

Afox. (2019, March 2). 5 Significant Benefits of Using Energy-Efficient Appliances. Astral Energy.https://www.astralenergyllc.com/5-significant-benefits-of-using-energy-efficient-appliances/

African Circular Business Alliance. (2023, March 12). Circular Economy Implementation Strategies for Sustainable Transportation. Www.linkedin.com. https://www.linkedin.com/pulse/circular-economy-implementation-strategies

Agrawal, S. K. (2021, July 19). ANN for Data Science | Basics Of Artificial Neural Network. Analytics Vidhya. https://www.analyticsvidhya.com/blog/2021/07/understanding-the-basics-of-artificial-neural-network-ann/

Ahammed, R. (2021, October 24). Climate change and global challenges. Times of India Blog. https://timesofindia.indiatimes.com/readersblog/redoanahammed/climate-change-and-global-challenges-38575/

Alwarritzi, M. J., NansekiTeruaki, Uenishi Yoshihiro, Widya. (2023, January 26). Japan guides the way on smart farming technology adoption. Asia Pathways. https://www.asiapathways-adbi.org/2023/01/japan-guides-the-way-on-smart-farming-technology-adoption/

American Chemical Society. (2021, November 3). Three ways to reduce the carbon footprint of food purchased by US households. ScienceDaily. https://www.sciencedaily.com/releases/2021/11/211103082607.htm#

Apoorv. (2011). India Map - Political, States and Capital Maps. Maps of India. https://www.mapsofindia.com/

Arneth, A., Sitch, S., Pongratz, J., D, S. B., Ciais, P., Poulter, B., D, B. A., Bondeau, A., Calle, L., P, C. L., Gasser, T., Fader, M., Friedlingstein, P., Kato, E., Li, W., Lindeskog, M., Nabel, J. E. M. S., Pugh, T. A. M., Robertson, E., and Viovy, N. (2017). Historical carbon dioxide emissions caused by landuse changes are possibly larger than assumed. Nature Geoscience, 10(2): 79–84.

Arora, N. K. (2019). Impact of climate change on agriculture production and its sustainable solutions. Environmental Sustainability, 2(2): 95–96.

Asseng, S. (2015). Uncertainties of Climate Change Impacts in Agriculture. Procedia Environmental Sciences, 29: 304.

Aydinalp, C., and Cresser, M. (2008). The Effects of Global Climate Change on Agriculture. American-Eurasian J. Agric. and Environ. Sci., 3(5): 672–676.

Bairagya, R. (2022). Productivity, Income, Livelihood and Ecology: The resilience analysis in Argo-ecosystem [Thesis].

Barrett, J. R. (2014). Climate Change Mitigation: Assessing Strategies that Offer Potential Human Health Benefits. Environmental Health Perspectives, 122(5).

Basuriya, K. (2023, June 1). Agriculture Sector. LMD. https://lmd.lk/agriculture-sector-20/

Baxi, S. (2023, May 15). Understanding Carbon Neutrality: A Key to Sustainable Future. Www.linkedin.com. https://www.linkedin.com/pulse/understanding-carbon-neutrality-key-sustainable-future-sahil-baxi-1f

BBC. (2019). Physical and human causes of climate change - Revision 3 - Higher Geography - BBC Bitesize. BBC Bitesize. https://www.bbc.co.uk/bitesize/guides/z3bbb9q/revision/3

BBC. (2023). Implications of climate change - Climate change - National 5 Geography Revision. BBC Bitesize. https://www.bbc.co.uk/bitesize/guides/z432pv4/revision/4

BBC Bitesize. (2012). Climate change - AQA - Revision 4 - GCSE Geography - BBC Bitesize. BBC Bitesize. https://www.bbc.co.uk/bitesize/guides/zx234j6/revision/4

Benton, T., Bieg, C., Harwatt, H., Pudasaini, R., and Wellesley, L. (2021). Food system impacts on biodiversity loss: Three levers for food system transformation in support of nature. Chatham House, Royal Institute of International Affairs.

Brandvoice, S. A. P. (2021, July 19). Fostering A Culture Of Innovation To Drive Growth. Forbes. https://www.forbes.com/sites/sap/2021/07/19/fostering-a-culture-of-innovation-to-drive-growth/?sh=3b8687b6235f

Brienen, Caldwell, L., Duchesne, L., Voelker, S., Barichivich, J., Baliva, M., Ceccantini, G., Filippo, D., Helama, S., M, L. G., Lopez, L., Piovesan, G., Schöngart, J., Villalba, R., and Gloor, E. (2020). Forest carbon sink neutralized by pervasive growthlife-spantrade-offs. Nature Communications, 11(1): 4241.

British Geological Survey. (2022). The greenhouse effect. British Geological Survey. https://www.bgs.ac.uk/discovering-geology/climate-change/how-does-the-greenhouse-effect-work/

Bruin, W. B. de, and Morgan, M. G. (2019). Reflections on an interdisciplinary collaboration to inform public understanding of climate change, mitigation, and impacts. Proceedings of the National Academy of Sciences, 116(16): 7676–7683.

Buis, A. (2020, March 10). How Climate Change May Be Impacting Storms Over Earth's Tropical Oceans. Climate Change: Vital Signs of the Planet. https://climate.nasa.gov/explore/ask-nasa-climate/2956/how-climate-change-may-be-impacting-storms-over-earths-tropical-oceans/

Bush, K. F., Luber, G., Kotha, S. R., Dhaliwal, R. S., Kapil, V., Pascual, M., Brown, D. G., Frumkin, H., Dhiman, R. C., Hess, J., Wilson, M. L., Balakrishnan, K., Eisenberg, J., Kaur, T., Rood, R., Batterman, S., Joseph, A., Gronlund, C. J., Agrawal, A., and Hu, H. (2011). Impacts of Climate Change on Public Health in India: Future Research Directions. Environmental Health Perspectives, 119(6): 765–770.

C-DAC. (2020). Sources of Energy. Vikaspedia.in. https://vikaspedia.in/energy/energy-basics/sources-of-energy

C. Alfred Jacobson. (1916). The Importance of Scientific Research to the Industries. Science, 44(1135): 456–459.

C2ES. (2018, October 30). Controlling Industrial Greenhouse Gas Emissions. Center for Climate and Energy Solutions. https://www.c2es.org/content/regulating-industrial-sector-carbon-emissions/

Campbellsville University. (2019, February 8). A Holistic View of Social Work. CU Online. https://online.campbellsville.edu/social-work/ecological-perspective-holistic-social-work/

Canzoneri, N. (2021, September 28). How Individual Actions Contribute to Collective Action for the Planet. Grow Ensemble. https://growensemble.com/individual-action/

Chandra, S. (2022). Happiness of Farmers: The Socio-ecological Interpretation and Strategy Implications [Thesis].

Chandra, S. (2022). Happiness of Farmers: The Socio-ecological Interpretation and Strategy Implications [Thesis].

Climate Change and Environment Strategy Consultation. (2019). Consult Cambridgeshire. https://consultcambs.uk.engagementhq.com/climate-strategy

Climate Change Science Adequately Reflected in the "Climate Change Sense of the Congress Resolution." (2003).

CSF. (2021). Communications Strategy FY21-23. The World Bank Group.

DISD/DPIDG. (2021, February 18). Leveraging digital technologies for social inclusion. Www.un.org. https://www.un.org/development/desa/dspd/2021/02/digital-technologies-for-social-inclusion/

EnglishClub. (2023). Climate change vocabulary. EnglishClub. Www.englishclub.com. https://www.englishclub.com/glossaries/climate-change.php

EPA. (2010). Climate Change and Ecosystems. United States Environmental Protection Agency.

European Climate Pact. (2022). Become an Ambassador or a "Friend of the Pact." Climate-Pact.europa.eu. https://climate-pact.europa.eu/ambassadors/become-ambassador-or-friend-pact_en

Farming First. (n.d.). Farmers Taking Action on Climate Change. Farming First. Retrieved June 4, 2023, from https://farmingfirst.org/farmers-taking-action-on-climate-change/

Fecht, S. (2021, February 25). How Exactly Does Carbon Dioxide Cause Global Warming? State of the Planet; Columbia Climate School. https://news.climate.columbia.edu/2021/02/25/carbon-dioxide-cause-global-warming/

Foster, S., and Elzinga, D. (2015, December). The Role of Fossil Fuels in a Sustainable Energy System. United Nations. https://www.un.org/en/chronicle/article/role-fossil-fuels-sustainable-energy-system

Friedrichsen, C. N., Daroub, S. H., Monroe, M. C., Stepp, J. R., and Wani, S. P. (2018). Mental Models of Soil Management for Food Security in Peri□Urban India. Urban Agriculture and Regional Food Systems, 3(1): 1–16.

GIZ. (2019). Innovative development planning for adaptation to climate change. Www.giz.de. https://www.giz.de/en/worldwide/58587.html

Global Climate Action. (2020). Climate Action Pathway: Land Use. United Nations Climate Change.

Google Scholar. (2023). Google Scholar. Google.com. https://scholar.google.com/

Gornall, J., Betts, R., Burke, E., Clark, R., Camp, J., Willett, K., and Wiltshire, A. (2010). Implications of climate change for agricultural productivity in the early twenty-first century. Philosophical Transactions of the Royal Society B: Biological Sciences, 365(1554): 2973–2989.

Grammarly. (2019). Write your best with Grammarly. Grammarly.com. https://www.grammarly.com

Gray, S., Chan, A., Clark, D., and Jordan, R. (2012). Modeling the integration of stakeholder knowledge in social–ecological decision-making: Benefits and limitations to knowledge diversity. Ecological Modelling, 229: 88–96.

Gray, S., Zanre, E., and Gray, S. (2012). Fuzzy Cognitive Maps as Representations of Mental Models and Group Beliefs.

Gray, S. A., Gray, S., Cox, L. J., and Henly-Shepard, S. (2013, January). Mental Modeler: A Fuzzy-Logic Cognitive Mapping Modeling Tool for Adaptive Environmental Management. 2013 46th Hawaii International Conference on System Sciences. https://doi.org/10.1109/hicss.2013.399

Gray, S. R. J., Gagnon, A. S., Gray, S. A., O'Dwyer, B., O'Mahony, C., Muir, D., Devoy, R. J. N., Falaleeva, M., and Gault, J. (2014). Are coastal managers detecting the problem?

Assessing stakeholder perception of climate vulnerability using Fuzzy Cognitive Mapping. Ocean and Coastal Management, 94: 74–89.

Grimm, N. B., Chapin, F. S., Bierwagen, B., Gonzalez, P., Groffman, P. M., Luo, Y., Melton, F., Nadelhoffer, K., Pairis, A., Raymond, P. A., Schimel, J., and Williamson, C. E. (2013). The impacts of climate change on ecosystem structure and function. Frontiers in Ecology and the Environment, 11(9): 474–482.

Guntukula, R., and Goyari, P. (2020). Climate Change Effects on the Crop Yield and Its Variability in Telangana, India. Studies in Microeconomics, 8(1): 119–148.

Hahn, L. (2020, May 5). How can communities reduce their carbon footprint? An introduction to community choice aggregation and community science. Program on Climate Change. https://pcc.uw.edu/blog/2020/05/05/how-can-communities-reduce-their-carbon-footprint-an-introduction-to-community-choice-aggregation-and-community-science/

Halbrendt, J., Gray, S. A., Crow, S., Radovich, T., Kimura, A. H., and Tamang, B. B. (2014). Differences in farmer and expert beliefs and the perceived impacts of conservation agriculture. Global Environmental Change, 28: 50–62.

Haringhata B.D.O., and InfoSky Solutions. (2022). B.D.O Desk. Www.haringhata.org. https://www.haringhata.org/bdo-desk.php

Hazelton, A. (2021, June 8). Why is scientific collaboration key? 4 experts explain. World Economic Forum. https://www.weforum.org/agenda/2021/06/4-views-on-why-scientific-collaboration-is-key-for-the-future/

Heal, G. M., and Kristrom, B. (2002). Uncertainty and Climate Change. SSRN Electronic Journal. https://doi.org/10.2139/ssrn.302399

Henly-Shepard, S., Gray, S. A., and Cox, L. J. (2015). The use of participatory modeling to promote social learning and facilitate community disaster planning. Environmental Science and Policy, 45: 109–122.

Hoffman, M., Lubell, M., and Hillis, V. (2014). Linking knowledge and action through mental models of sustainable agriculture. Proceedings of the National Academy of Sciences, 111(36): 13016–13021.

Howarth, C., and Painter, J. (2016). Exploring the science–policy interface on climate change: The role of the IPCC in informing local decisionmaking in the UK. Palgrave Communications, 2(1).

IAEA. (2016, April 13). Development of smart agriculture practices. Iaea.org. https://www.iaea.org/topics/development-of-smart-agricultural-practices

IBEF. (2021, June 25). India UK Initiative To Promote Industrial Energy Efficiency. India Brand Equity Foundation. https://www.ibef.org/blogs/the-12th-chief-energy-ministerial-india-uk-initiative-to-promote-industrial-energy-efficiency

IEA. (2021). Carbon capture, utilisation and storage. IEA. https://www.iea.org/fuels-and-technologies/carbon-capture-utilisation-and-storage

ILO. (n.d.). Disaster preparedness and response in Pacific island countries. Www.ilo.org. Retrieved May 24, 2023, from https://www.ilo.org/suva/areas-of-work/disaster-mitigation-and-recovery/lang--en/index.htm

Ingenium. (n.d.). Let's Talk Energy - Climate Change Mitigation and Adaptation. Energy. techno-Science.ca. Retrieved June 18, 2023, from https://energy.techno-science.ca/en/climate-101-climate-change-mitigation-and-adaptation.php

IPCC. (2011). Renewable Energy Sources and Climate Change Mitigation. Ipcc.ch; IPCC. https://www.ipcc.ch/report/renewable-energy-sources-and-climate-change-mitigation/

IPCC. (2019). Working Group III — IPCC. Ipcc.ch; IPCC. https://www.ipcc.ch/working-group/wg3/

IPCC. (2022, February 27). Climate Change 2022: Impacts, Adaptation and Vulnerability. Www.ipcc.ch. https://www.ipcc.ch/report/ar6/wg2/

IPCC. (2022a). Global Warming of 1.5°C. Annex I: Glossary (pp. 541–562). https://doi.org/10.1017/9781009157940.008

IPCC. (2022b). IPCC — Intergovernmental Panel on Climate Change. Ipcc.ch; IPCC. https://www.ipcc.ch/

Kamali, B., Lorite, I. J., Webber, H. A., Rezaei, E. E., Gabaldon-Leal, C., Nendel, C., Siebert, S., Ramirez-Cuesta, J. M., Ewert, F., and Ojeda, J. J. (2022). Uncertainty in climate change impact studies for irrigated maize cropping systems in southern Spain. Scientific Reports, 12(1).

Ki'Amber Thompson. (2019, March 22). Five Ways Our Ocean Contributes to Freshwater Access, or Lack Thereof. Ocean Conservancy. https://oceanconservancy.org/blog/2019/03/22/five-ways-ocean-contributes-freshwater-access-lack-thereof/

Koch, I. M. (2022, September 19). Council Post: How Companies Can Embrace Conscious Consumption. Forbes. https://www.forbes.com/sites/forbesbusinesscouncil/2022/09/19/how-companies-can-embrace-conscious-consumption/?sh=3df497686bf1

Kundu, T. (2022). Entrepreneurial Communication: The Process, Factors and Impact in Agricultural Enterprises [Thesis].

Kunreuther, H., Gupta, S., Bosetti, V., Cooke, R., Dutt, V., Ha-Duong, M., Held, H., Llanes-Regueiro, J., Patt, A., Shittu, E., and Weber, E. (2014). Integrated Risk and Uncertainty Assessment of Climate Change Response Policies. Cambridge University Press.

Lamaoui, M., Jemo, M., Datla, R., and Bekkaoui, F. (2018). Heat and Drought Stresses in Crops and Approaches for Their Mitigation. Frontiers in Chemistry, 6.

Lattner, J. (2021, March 12). Renewable Power for Carbon Dioxide Mitigation. Energy Central. https://energycentral.com/news/renewable-power-carbon-dioxide-mitigation

Leong, C. (2016). Resilience to climate change events: The paradox of water (In) security. Sustainable Cities and Society, 27: 439–447.

Levy, M. A., Lubell, M. N., and McRoberts, N. (2018). The structure of mental models of sustainable agriculture. Nature Sustainability, 1(8): 413–420.

Lewandowski, S., and Ullrich, A. (2023). Measures to reduce corporate GHG emissions: A review based taxonomy and survey based cluster analysis of their application and perceived effectiveness. Journal of Environmental Management, 325.

Mahato, A. (2014). Climate Change and its Impact on Agriculture. International Journal of Scientific and Research Publications, 4(4).

Maizland, L. (2022, November 4). Global Climate Agreements: Successes and Failures. Council on Foreign Relations. https://www.cfr.org/backgrounder/paris-global-climate-change-agreements

Malcolm, J. R., and Pitelka, L. F. (2000). Ecosystems and Global climate change: A Review of Potential Impacts on U.S. Terrestrial Ecosystems and Biodiversity. Pew Center on Global Climate Change.

Malhi, Y., Franklin, J., Seddon, N., Solan, M., Turner, M. G., Field, C. B., and Knowlton, N. (2020). Climate change and ecosystems: threats, opportunities and solutions. Philosophical Transactions of the Royal Society B: Biological Sciences, 375(1794).

Mandal, K. G., Thakur, A. K., Mohanty, R. K., Mishra, A. K., Sinha, S., and Biswas, B. (2022). Policy perspectives on agricultural water management and associated technologies suitable for different agro-climatic zones of West Bengal, India. Current Science, 122(4): 386.

Mastrandrea, M., Field, C., Stocker, T., Ebi, K., Frame, D., Held, H., Kriegler, E., Mach, K., Matschoss, P., Plattner, G.-K., Yohe, G., and Zwiers, F. (2010). Guidance Note for Lead

Authors of the IPCC Fifth Assessment Report on Consistent Treatment of Uncertainties. Intergovernmental Panel on Climate Change (IPCC).

Mehryar, S. (2022, September 12). What is the difference between climate change adaptation and resilience? Grantham Research Institute on Climate Change and the Environment. https://www.lse.ac.uk/granthaminstitute/explainers/what-is-the-difference-between-climate-change-adaptation-and-resilience/

Mehta, L., Adam, H., and Srivastava, S. (2021). The Politics of Climate Change and Uncertainty in India. Routledge. https://doi.org/10.4324/9781003257585

Mehta, L., Adam, H. N., and Srivastava, S. (2019). Unpacking uncertainty and climate change from "above" and "below." Regional Environmental Change, 19(6): 1529–1532.

Mehta, L., Srivastava, S., Adam, H. N., Alankar, Bose, S., Ghosh, U., and Kumar, V. V. (2019). Climate change and uncertainty from "above" and "below": perspectives from India. Regional Environmental Change, 19(6): 1533–1547.

Mendeley - Reference Management Software and Researcher Network. (2018). Mendeley.com. https://www.mendeley.com/

Mishra, D., and Sahu, N. C. (2014). Economic Impact of Climate Change on Agriculture Sector of Coastal Odisha. APCBEE Procedia, 10: 241–245.

Modelling and projections. (2021, March 22). Www.climatechangeinaustralia.gov.au. https://www.climatechangeinaustralia.gov.au/en/learning-support/modelling-and-projections/

Moore, F., Uris Lantz, C. B., and Hertel, T. (2017). Economic impacts of climate change on agriculture: a comparison of process-based and statistical yield models. Environ. Res. Lett., 12.

MyBib. (2018). MyBib. MyBib. https://mybib.com

NASA. (2018, September 24). Sea level, NASA Global Climate Change. Climate Change: Vital Signs of the Planet; NASA. https://climate.nasa.gov/vital-signs/sea-level/

NASA. (2019). What is the greenhouse effect? Climate Change: Vital Signs of the Planet; NASA. https://climate.nasa.gov/faq/19/what-is-the-greenhouse-effect/

Nastis, S. A., Michailidis, A., and Chatzitheodoridis, F. (2012). Climate change and agricultural productivity. African Journal of Agricultural Research, 7(35): 4885–4893.

Natalie. (2022, June 2). Conceptual and Theoretical Frameworks for Thesis Studies: What you must know. Enago.com. https://www.enago.com/thesis-editing/blog/conceptual-and-theoretical-frameworks-for-thesis-studies-what-you-must-know

National Oceanic and Atmospheric Administration. (2008). National Oceanic and Atmospheric Administration. Noaa.gov. https://www.noaa.gov/

National Security Council. (2021). U.S. Strategy for Addressing the Root Causes of Migration in Central America. The White House.

NOAA. (2021a, August 13). Climate change impacts. Www.noaa.gov; NOAA. https://www.noaa.gov/education/resource-collections/climate/climate-change-impacts

NOAA. (2021b, December). Satellite and space resources. Www.noaa.gov. https://www.noaa.gov/agu-2021/NOAA-agu-focus-areas/satellite-and-space-resources

Noleppa, S., and Agripol. (2013). Economic approaches for assessing climate change adaptation options under uncertainty Excel tools for Cost-Benefit and Multi-Criteria Analysis Competence Center for Climate Change. Deutsche Gesellschaftfür Internationale Zusammenarbeit (GIZ) GmbH.

Nyaki, A., Gray, S. A., Lepczyk, C. A., Skibins, J. C., and Rentsch, D. (2014). Local-Scale Dynamics and Local Drivers of Bushmeat Trade. Conservation Biology, 28(5): 1403–1414.

Nye, J. (2010). Climate Change and Its Effects on Ecosystems, Habitats and Biota. Gulf of Maine Council on the Marine Environment.

Nápoles, G., Espinosa, M. L., Grau, I., and Vanhoof, K. (2018). FCM Expert: Software Tool for Scenario Analysis and Pattern Classification Based on Fuzzy Cognitive Maps. International Journal on Artificial Intelligence Tools, 27(07).

OECD. (2016). The OECD supporting action on climate change. OECD. http://oe.cd/climate

Okello, J., Zhou, Y., Barker, I., and Schulte-Geldermann, E. (2018). Motivations and Mental Models Associated with Smallholder Farmers' Adoption of Improved Agricultural Technology: Evidence from Use of Quality Seed Potato in Kenya. The European Journal of Development Research, 31(2): 271–292.

OpenAI. (2023). ChatGPT. Chat.openai.com; OpenAI. https://chat.openai.com/

Ortiz, A. M. D., Outhwaite, C. L., Dalin, C., and Newbold, T. (2021). A review of the interactions between biodiversity, agriculture, climate change, and international trade: research and policy priorities. One Earth, 4(1): 88–101.

Pal, D. (2019, November 24). Evidence-based policy-making is the way forward. The Hindu. https://www.thehindu.com/opinion/open-page/evidence-based-policy-making-is-the-way-forward/article30061181.ece

Pallavi, D. M. (2023, May 12). The Climate Imperative: Navigating a Sustainable Path Forward. Medium.com; ILLUMINATION-Curated. https://medium.com/illumination-curated/the-climate-imperative-navigating-a-sustainable-path-forward-55f362d10889

Pandit, P., Roy, S., Chatterjee, D., Islam, S., and Mohapatra, U. (2019). Application of Statistical Analysis Software and Packages for Research Data of Agriculture and Allied Sectors - a Short Communication. Bull. Env. Pharmacol. Life Sci, 8(11): 110–113.

Panse, V., and Sukhatme, P. (1967). Statistical Methods for Agricultural Workers. Indian Council of Agricultural Research, 381p.

Patra, S. (2022). Uncertainty and Chaos in farming: An interpretation on production, livelihood, and income [Thesis].

Patra, S., Acharya, S. K., Chandra, S., Biswas, A., and Haque, M. (2022). Uncertainty and Vulnerability: On the pathway of modernizing Indian Agriculture. Journal of Community Mobilization and Sustainable Development, 3(Seminar Special Issue): 1008–1014.

PIB Delhi. (2022, December 16). Impact of Climate Change on Agriculture. Pib.gov.in. https://pib.gov.in/PressReleaseIframePage.aspx?PRID=1884236

Pineo, H., Zimmermann, N., and Davies, M. (2020). Integrating health into the complex urban planning policy and decision-making context: a systems thinking analysis. Palgrave Communications, 6(1).

PNAS. (2016, August 16). Climate change and irrigation water. Www.pnas.org. https://doi.org/10.1073/pc.10122283

Quillbot. (2019). QuillBot | Free Paraphrasing Tool - Best Article Rewriter. Quillbot.com. https://quillbot.com/

Ray, G. L., and Mondal, S. (2011). Research methods in social sciences and extension education. Kalyani Publishers, 330p.

Research Gate. (2019). ResearchGate | share and discover research. ResearchGate; ResearchGate. https://www.researchgate.net/

Reynolds, T. W., Bostrom, A., Read, D., and Morgan, M. G. (2010). Now What Do People Know About Global Climate Change? Survey Studies of Educated Laypeople. Risk Analysis, 30(10): 1520–1538.

Royalty Free Stock Photos, Illustrations, Vector Art and Video Clips - Getty Images. (n.d.). Www.gettyimages.in. https://www.gettyimages.in/

Sci-Hub: removing barriers in the way of science. (2011). Sci-Hub.se. https://sci-hub.se/

Sharifi, A. (2020). Tradeoffs and conflicts between urban climate change mitigation and adaptation measures: A literature review. Journal of Cleaner Production, 276.

Shen, W., Han, W., and Wallington, T. J. (2014). Current and Future Greenhouse Gas Emissions Associated with Electricity Generation in China: Implications for Electric Vehicles. Environmental Science and Technology, 48(12): 7069–7075.

Sherwood, S. C., Webb, M. J., Annan, J. D., Armour, K. C., Forster, P. M., Hargreaves, J. C., Hegerl, G., Klein, S. A., Marvel, K. D., Rohling, E. J., Watanabe, M., Andrews, T., Braconnot, P., Bretherton, C. S., Foster, G. L., Hausfather, Z., Heydt, A. S., Knutti, R., Mauritsen, T., and Norris, J. R. (2020). An Assessment of Earth's Climate Sensitivity Using Multiple Lines of Evidence. Reviews of Geophysics, 58(4).

Shi, Z., Huang, H., Wu, Y., Chiu, Y.-H., and Qin, S. (2020). Climate Change Impacts on Agricultural Production and Crop Disaster Area in China. International Journal of Environmental Research and Public Health, 17(13): 4792.

Shipman, M. (2020, August 26). Climate Change, Crop Yields and Risk Management for Farmers. NC State News. https://news.ncsu.edu/2020/08/climate-change-crop-yields-and-risk-management-for-farmers/

Shoaib, S. A., Khan, M. Z. K., Sultana, N., and Mahmood, T. H. (2021). Quantifying Uncertainty in Food Security Modeling. Agriculture, 11(1): 33.

Simon, C. (2021, September 21). Melting of polar ice shifting Earth itself, not just sea levels. Harvard Gazette. https://news.harvard.edu/gazette/story/2021/09/melting-of-polar-ice-shifting-earth-itself-not-just-sea-levels/

Singh, R. K., and Chaudhary, B. D. (1967). Biometrical Methods in Quantitative Genetic Analysis. Kalyani Publishers, 480p.

Spash, Clive L. (2011). Social Ecological Economics: Understanding the Past to See the Future. American Journal of Economics and Sociology, 70(2): 340–375.

Squires, V. R., and Gaur, M. K. (2020). Agricultural Productivity and Food Security. In V. R. Squires and M. K. Gaur (Eds.), Food Security and Land Use Change under Conditions of Climatic Variability: A Multidimensional Perspective. Springer International Publishing. pp. 33–48.

Srinivasa Rao, Ch., Prasad, R., and Mohapatra, T. (2019). Climate Change and Indian Agriculture: Impacts, Coping Strategies, Programmes and Policy (p. 25). Indian Council of Agricultural Research.

Stecula, D. A., and Merkley, E. (2019). Framing Climate Change: Economics, Ideology, and Uncertainty in American News Media Content From 1988 to 2014. Frontiers in Communication, 4.

Tabassum, W. (2023, April 11). The threat of global warming: Understanding its causes and effects on our world. Indiatimes. https://timesofindia.indiatimes.com/readersblog/ajourneytotheseashore/the-threat-of-global-warming-understanding-its-causes-and-effects-on-our-world-52465/

Tech Fusion. (2021). The Butterfly Graph Quantum In Effect and Chaos Theory! In TechFusionNews. https://techfusion24.com/the-butterfly-graph-quantum-in-effect-chaos-theory/

Ten Powerful Mental Models: How To Make Better Decisions And Gain Clarity In Thinking. (n.d.). Play for Thoughts. Retrieved June 17, 2023, from https://www.playforthoughts.com/blog/powerful-mental-models

Thakur, B. R., Sharma, D. D., and Mohan, P. (2014). Dynamics of Agricultural Production. Livelihood Security in Northwestern Himalaya, 67–82. https://doi.org/10.1007/978-4-431-54868-3_6

Thomas, C., Ohlemüller, R., Anderson, B., Hickler, T., Miller, P., Sykes, M., and Williams, J. (2008). Exporting the ecological effects of climate change Developed and developing countries will suffer the consequences of climate change, but differ in both their responsibility and how badly it will affect their ecosystems. EMBO Reports, 9.

Turgeon, A., and Morse, E. (2022, July 28). Solar Energy. National Geographic Society. Education.nationalgeographic.org; National Geographic Society. https://education.nationalgeographic.org/resource/solar-energy/

Turner, B. L., Wuellner, M., Nichols, T., and Gates, R. (2014). Dueling Land Ethics: Uncovering Agricultural Stakeholder Mental Models to Better Understand Recent Land Use Conversion. Journal of Agricultural and Environmental Ethics, 27(5): 831–856.

UCAR. (2021). The Water Cycle and Climate Change | UCAR Center for Science Education. Scied.ucar.edu. https://scied.ucar.edu/learning-zone/climate-change-impacts/water-cycle-climate-change

UCDAVIS. (n.d.). Climate Change and Resilience at the Community Level. Online.ucdavis.edu. Retrieved May 24, 2023, from https://online.ucdavis.edu/course/climate-change-and-resilience-community-level

United Nations. (2021). Education is key to addressing climate change. United Nations. https://www.un.org/en/climatechange/climate-solutions/education-key-addressing-climate-change

United Nations. (2022a). Food and Climate Change: Healthy diets for a healthier planet. United Nations. https://www.un.org/en/climatechange/science/climate-issues/food

United Nations. (2022b). What is Climate Change? Climate Action; United Nations. https://www.un.org/en/climatechange/what-is-climate-change

United States Environmental Protection Agency. (2017, November 2). Greenhouse Gases | US EPA. US EPA. https://www.epa.gov/report-environment/greenhouse-gases

UN Secretary General. (2015). Strengthening and coordinating United Nations rule of law activities. In www.un.org. General Assembly of the United Nations. https://www.un.org/en/ga/sixth/70/rule_of_law.shtml

US EPA, OW. (2015, October). Coastal Resiliency. US EPA. https://www.epa.gov/green-infrastructure/coastal-resiliency

US EPA. (2022, May 16). Overview of Greenhouse Gases. US EPA; United States Environmental Protection Agency. https://www.epa.gov/ghgemissions/overview-greenhouse-gases

Van Hulst, F., Ellis, R., Prager, K., and Msika, J. (2020). Using co-constructed mental models to understand stakeholder perspectives on agro-ecology. International Journal of Agricultural Sustainability, 18(2): 172–195.

Vedantu. (2019, April 10). Effects of Burning Fossil Fuels. Vedantu.com; Vedantu. https://www.vedantu.com/chemistry/effects-of-burning-fossil-fuels

Waldman, K. B., Todd, P. M., Omar, S., Blekking, J. P., Giroux, S. A., Attari, S. Z., Baylis, K., and Evans, T. P. (2020). Agricultural decision making and climate uncertainty in developing countries. Environmental Research Letters, 15(11).

Walters, J. P., Archer, D. W., Sassenrath, G. F., Hendrickson, J. R., Hanson, J. D., Halloran, J. M., Vadas, P., and Alarcon, V. J. (2016). Exploring agricultural production systems and their fundamental components with system dynamics modelling. Ecological Modelling, 333: 51–65.

Warsaw, P., Archambault, S., He, A., and Miller, S. (2021). The Economic, Social, and Environmental Impacts of Farmers Markets: Recent Evidence from the US. Sustainability, 13(6): 3423.

WEF. (2021, July 23). How to build long-term resilience in vulnerable communities. World Economic Forum. https://www.weforum.org/agenda/2021/07/how-to-build-long-term-resilience-in-vulnerable-communities/

Weiskopf, S. R., Rubenstein, M. A., Crozier, L. G., Gaichas, S., Griffis, R., Halofsky, J. E., Hyde, K. J. W., Morelli, T. L., Morisette, J. T., Muñoz, R. C., Pershing, A. J., Peterson, D. L., Poudel, R., Staudinger, M. D., Sutton-Grier, A. E., Thompson, L., Vose, J., Weltzin, J. F., and Whyte, K. P. (2020). Climate change effects on biodiversity, ecosystems, ecosystem services, and natural resource management in the United States. Science of the Total Environment, 733: 137-782.

Wikipedia. (2022, July 11). Effects of climate change on ecosystems. Wikipedia. https://en.wikipedia.org/wiki/Effects_of_climate_change_on_ecosystems

Wikipedia. (2023, May 15). Greenhouse gas monitoring. Wikipedia. https://en.wikipedia.org/wiki/Greenhouse_gas_monitoring

Wikipedia Contributors. (2018, December 31). Efficient energy use. Wikipedia; Wikimedia Foundation. https://en.wikipedia.org/wiki/Efficient_energy_use

Wikipedia Contributors. (2019, April 12). Mental model. Wikipedia; Wikimedia Foundation. https://en.wikipedia.org/wiki/Mental_model

Wikipedia Contributors. (2019, January 26). Environmental impact of agriculture. Wikipedia; Wikimedia Foundation. https://en.wikipedia.org/wiki/Environmental_impact_of_agriculture

WION. (2018, November 6). Private vehicles contribute nearly 40% to air pollution in Delhi, action needed for minimising their use: CSE. WION. https://www.wionews.com/india-news/private-vehicles-contribute-nearly-40-to-air-pollution-in-delhi-action-needed-for-minimizing-their-use-cse-175689

WMO. (2015, November 27). About us. World Meteorological Organization. https://public.wmo.int/en/about-us

WMO. (2020, October 13). State of Climate Services 2020 Report: Move from Early Warnings to Early Action. World Meteorological Organization. https://public.wmo.int/en/media/press-release/state-of-climate-services-2020-report-move-from-early-warnings-early-action

World 101. (2019). The Greenhouse Effect. World101 from the Council on Foreign Relations. https://world101.cfr.org/global-era-issues/climate-change/greenhouse-effect

World Meteorological Organization. (2019). World Meteorological Organization. World Meteorological Organization. https://public.wmo.int/en

Öhlmér, B. (1998). Understanding farmers' decision making processes and improving managerial assistance. Agricultural Economics, 18(3): 273–290.

Images of Data Collection
1: Bhomra Beal survey

Images of Data Collection
2: Haringhata ADA Office

Images of Data Collection 3: Kastadanga-I GP

Images of Data Collection 4: Kastadanga SKUS Ltd.

Images of Data Collection 5: One Contact Farmer-respondent (randomly selected)

Index